Louis H. Gibson

Beautiful Houses

A study in house-building

Louis H. Gibson

Beautiful Houses
A study in house-building

ISBN/EAN: 9783337419776

Printed in Europe, USA, Canada, Australia, Japan

Cover: Foto ©berggeist007 / pixelio.de

More available books at **www.hansebooks.com**

A Study in House-building

FOREIGN EXAMPLES IN DOMESTIC AR-
CHITECTURE — A COLLECTION OF
AMERICAN HOUSE PLANS — MATERIALS
AND DETAILS FOR THE ARTISTIC
HOUSE - BUILDER — THE ARCHITECT

BY

LOUIS H. GIBSON, ARCHITECT

Author of "Convenient Houses"

PREFACE.

ART, as applied to homes, furniture, the utensils of life,—in a word, the democratic art,—joins hands with the exclusive art of the easel.

For the sake of clearness and system, this book is distinctly divided into sections. The first, *House-building and Art*, shows that all building may be artistic, that there is no good reason for ugliness, that it costs no more to make an artistic detail than to make an ugly one. There need be no more material and labor in one than in the other. It is a question of direction by a trained artist.

Examples of *The World's Homes* are selected from various countries to show how each region has worked out its own solution of the problem of domestic art. The work of France, England, Germany, Switzerland, and our own country during the Old Colonial period has been widely different as to method, and reasonably uniform as to success. These examples indicate that we may hope to do our own work in our own way, preserve our own character, and develop a successful American architecture.

The section under headings of *Some House Plans* and *Materials and Details* is an attempt at the practical working out of the general ideas involved in this book.

The last division is addressed to the client. It explains the artistic and business sides of the profession. At the same time it shows what the client may do to assist in a better architectural development.

In a former volume the subject of "Convenient Houses" was treated, and a large number of small plans given.

The various works from which illustrations have been taken are indicated in the text. With one exception I have been able to credit the work of other architects whose work I have used. My efforts to obtain the name of the architect of the house designated "In California" were unsuccessful.

The cover of this book was designed by my brother, David Gibson, who has also helped me in the preparation of the illustrations.

I am indebted to my friend William Forsyth, the artist, for criticism and suggestion.

Mr. Louis H. Sullivan, the architect, whose work I hold in great esteem, furnished me a number of the decorative designs used for tail-pieces.

My wife has edited and criticised the manuscript, and during the five years spent in the selection of material, has given her invaluable help.

LOUIS H. GIBSON.

INDIANAPOLIS, IND., May, 1895.

CONTENTS.

HOUSE-BUILDING AN ART.

CHAPTER I.

UGLY HOUSES. — UNEDUCATED ARCHITECTS. — THE ARTIST HOUSE-BUILDER. — THE MOST EXPENSIVE HOUSES NOT NECESSARILY THE MOST ARTISTIC. — THE ARTISTIC FAILURE OF GOVERNMENT BUILDINGS. — COST NEVER MEASURES THE ARTISTIC. — ORNAMENT AND LABOR. — ARTISTIC DETAILS COST NO MORE THAN UGLY ONES. — COMMERCIAL VALUE OF THE ARTISTIC 9

CHAPTER II.

DEVELOPMENT OF ART IN BUILDING. — THE PRIMITIVE HOUSE. — FIRST PRINCIPLES. — THE GREEK TEMPLE AND THE INDIAN HUT. — THE FUNCTIONAL AND THE ARTISTIC. — ROMAN METHODS OF BUILDING. — ACADEMIC ARCHITECTURE. — THE OLD ROMAN AND THE OLD COLONIAL. — ROMANESQUE ARCHITECTURE. — GOTHIC ARCHITECTURE. — EMOTIONAL EXPRESSION IN BUILDING. — DECLINE OF THE GOTHIC. — THE RENAISSANCE. — MODERN ARCHITECTURE OF EUROPE. — CHARACTERISTICS OF MODERN AMERICAN ARCHITECTURE. — AMERICAN STRUCTURAL FORMS. — AMERICAN METHODS OF LIVING AS RELATED TO ARCHITECTURE. — RELATION OF PRECEDENT TO FUTURE DEVELOPMENT . . 17

THE WORLD'S HOMES.

CHAPTER III.

FRENCH DOMESTIC ARCHITECTURE. — NATIONAL BUILDING METHODS. — TWELFTH-CENTURY BUILDING. — THE MUSICIAN'S HOUSE AT RHEIMS. — FLOOR PLANS OF DOMESTIC STRUCTURES. — PICTURESQUE STAIR TOWERS. — THE HOUSE OF JACQUES COEUR. — AN EARLY FARM-HOUSE. — HALF-TIMBER ARCHITECTURE OF THE TWELFTH CENTURY. — LATER DEVELOPMENTS IN WOODEN ARCHITECTURE. — OUR USE OF FRENCH EXAMPLES. — FROM LISSIEUX 32

CHAPTER IV.

BRITTANY. — THE PAINTER'S COUNTRY. — FEW OFFENSIVE BUILDINGS. — CHARACTERISTICS OF THE PEOPLE. — THE ARCHITECTURE, THE SCENERY, AND THE PEOPLE HARMONIOUS. — BRETON COSTUMES. — FURNITURE. — EXTRAVAGANTLY PICTURESQUE. — MALESTROIT. — A CHIMNEY-PIECE AT MORLAIX. — BEDS 48

CHAPTER V.

FRENCH CHÂTEAUX. — MILITARY STRONGHOLDS. — PRINCIPLES GOVERNING THEIR CONSTRUCTION. — HISTORY. — PLAN OF COURCY. — MILITARY BUILDING AND ART. — A BIRD'S-EYE VIEW. — CHÂTEAU OF JOSSELIN. — INTERIOR VIEWS. — PIERREFONDS. — CHAUMONT. — A VIEW OF CHENONCEAU. — AZAY-LE-RIDEAU. — THE SOCIAL CHÂTEAUX 60

CHAPTER VI.

ENGLISH DOMESTIC ARCHITECTURE. — RELATION OF MODERN TO EARLY WORK. — SELECTIONS FROM HAMERTON'S PORTFOLIO. — DOMESTIC BUILDINGS OF THE SIXTEENTH AND SEVENTEENTH CENTURIES. — PICTURESQUE DETAILS. — FROM THE GOTHIC TO THE RENAISSANCE. — SIX GABLES. — DECORATIVE WOODWORK. — PUGIN'S GABLES. — THE SPIRIT OF DOMESTICITY. — WALTER SCOTT'S HOME. — ONE OF NORMAN SHAW'S INTERIORS 90

CHAPTER VII.

MODERN ARCHITECTURE OF GERMANY. — DRESDEN AND THE ROCOCO. — NORTH OF THE HARTZ MOUNTAINS. — THE OLD ROSE-TREE AT HILDESHEIM. — AN ARCHITECTURE FULL OF FINE SENTIMENT. — IMAGINATIVE ARCHITECTURE. — TILE ROOFS. — CLEAN PAVEMENTS. — DOORWAYS. — NATURAL METHODS . . . 108

CHAPTER VIII.

SWISS ARCHITECTURE. — RELATION OF SCENERY TO A NATION'S BUILDING. — NATURAL DOMESTIC EXPRESSIONS. — ARTISTIC FORMS. — NATURAL COLORING. — VARYING METHODS. — CHARACTERISTIC ILLUSTRATIONS . 123

CHAPTER IX.

OLD COLONIAL ARCHITECTURE. — A CLASSIC DEVELOPMENT. — CHARACTERISTIC NEW ENGLAND ARCHITECTURE. — LUXURIOUS CHARACTER OF THE OLD COLONIAL IN THE SOUTH. — THE BEST AMERICAN EXPRESSION OF DOMESTICITY. — ILLUSTRATIONS . 127

SOME HOUSE PLANS.

CHAPTER X.

A BRICK HOUSE WITH A STONE FOUNDATION. — FOR A NARROW LOT. — THE PROBLEM OF LIGHTING. — A NEW CLOSET. — RELATION OF THE EXTERIOR TO THE LOCATION. — THE DORMERS. — THE INSIDE FINISH. — MANTELS 135

CHAPTER XI.

A CENTRE-HALL PLAN. — FRAME BUILDING. — A LITTLE ROOM FOR CLOAKS AND WRAPS. — DECORATIVE FORMS. — INTERIOR PHOTOGRAPHS. — EXTERNAL DETAILS. — GREEK MOULDINGS . . . 144

CHAPTER XII.

A WIDE CENTRAL HALL, OPEN AT EACH END. — LARGE ROOMS. — A PICTURESQUE STAIRWAY. — COLOR SCHEMES IN DECORATION. — DESCRIPTION OF FLOOR PLAN 156

CHAPTER XIII.

A FINE LOCATION. — A RIVER FRONT. — PICTURESQUE STAIR-HALL. — A SMOKING-ROOM UNDER THE BALCONY. — WOOD CEILING. — DINING-ROOM FINISH. — KITCHEN DETAILS. — DOORS AND CASINGS. — GREEK PROFILES. — THE LOCATION PLAN 169

CHAPTER XIV.

A STABLE. — A WATER-TOWER. — PUMPS — A FLOOR PLAN. — INTERIOR DETAILS. — DRAINAGE OF THE FLOOR AND STALLS. — AN ODORLESS MANURE-BIN. — STALL CONSTRUCTION. — THE MAN'S ROOM 178

CHAPTER XV.

THE HOUSEKEEPER AND THE FLOOR PLAN. — THE SOBER-MINDED CLIENT. — THE ONE WITH PICTURESQUE TASTES. — SOUTH GERMAN ARCHITECTURE 184

CHAPTER XVI.

IN A FOREST. — SENTIMENT WHICH CONTROLS THE EXTERIOR. — A WIDE HALL. — THE MUSIC-ROOM. — THE END OF THE DINING-ROOM. — THE SIDEBOARD. — FLUSH-PANEL DOORS. — A SCREEN. — NATURAL-WOOD FINISH FOR THE EXTERIOR. — ROMAN BRICK. — BYZANTINE DETAIL . 190

CHAPTER XVII.

A LARGE NUMBER OF ROOMS. — BRICK AND STONE. — RENAISSANCE FORMS. — PICTURESQUE ROOF. — THE GABLES 200

CHAPTER XVIII.

STONE HOUSES. — A HOUSE FOR A WIDE LOT. — A PLAN WITH LIVING-ROOMS TO THE SOUTH. — THE FRONT. — FIFTEENTH CENTURY 205

CHAPTER XIX.

BUILDING FOR INVESTMENT. — DEVELOPMENT OF AN INVESTMENT BUILDING. — IMPROVEMENT OF RENTAL STRUCTURES. — RAPID TRANSIT AND RENTAL PROPERTY. — ONE-ROOM APARTMENTS IN GLASGOW. — LASTING QUALITIES OF FRAME HOUSES. — THE REPAIR ACCOUNT. — THE DOUBLE HOUSE. — BRICK HOUSES. — A SMALL FRAME HOUSE. — RANGES. — THE BETTER CLASS OF TENANTS. — ROOMS OVER STORES. — FRENCH AND ENGLISH FLATS. — A SMALL KITCHEN . 211

CHAPTER XX.

PERSONAL CHARACTER EXPRESSED IN HOUSE-BUILDING. — A HOUSE FOR THE BRIGHT, CHEERY LITTLE WOMAN. — FOR THE EXACT, DELIBERATE BUSINESS MAN. — THE PICTURESQUE CHARACTER. — LEGITIMATE VARIATION OF DESIGN INFLUENCED BY PERSONALITY. — A MISCELLANEOUS COLLECTION OF HOUSE PLANS 231

MATERIALS AND DETAILS.

CHAPTER XXI.

SHINGLE-HOUSES. — THE PROPER SURROUNDINGS. — THE STAINS OF TIME. — ARTIFICIAL STAIN. — EXAMPLES. — SLATE WALLS . . 242

CHAPTER XXII.

FIREPLACES AND MANTELS. — HISTORY. — MANTELS OF THE RENAISSANCE. — DUTCH MANTELS. — MODERN MANTELS. — CHARACTER IN MANTELS. — TILE FACINGS. — ONYX AND BRICK 253

CHAPTER XXIII.

DOORS. — THE DEFENSIVE. — HOSPITALITY. — MATERIAL. — FOREIGN EXAMPLES. — DOMESTIC DOORS 273

CHAPTER XXIV.

STAIRS. — FOREIGN EXAMPLES. — BRITTANY. — FRANCE. — HOLLAND. — GERMANY. — BROAD LANDINGS. — OLD COLONIAL STAIRWAYS. — IRON RAILINGS 282

CHAPTER XXV.

FURNITURE. — ARCHITECTS' DESIGNS. — SIDEBOARDS. — BOOKCASES. — SEATS. — LOUNGES. — SCREENS. — GRILLES 293

CHAPTER XXVI.

WALLS AND CEILINGS. — PUBLIC JUDGMENT. — THE MISLEADING INFLUENCE OF LARGE EXPENDITURES. — PERMANENCY IN DECORATION. — WALL PAPERS. — FRESCO. — JUTE. — DENIM. — SILK 302

CHAPTER XXVII.

MATERIALS. — KINDS OF WOOD. — MOULDINGS IN WOOD. — PLAIN SURFACES. — THE WOOD SCREEN AT AMIENS. — STAINING OF WOOD. — WOOD FINISHES. — WOOD FLOORS. — WORKMANSHIP 310

CHAPTER XXVIII.

THE ARTIST BLACKSMITH. — EARLY WORK. — HINGES. - LOCKS. — NAIL-HEADS 318

CHAPTER XXIX.

GLASS. — DECORATIVE USES. — RECENT FAILURES IN GLASS WORK. — ARTISTS' WORK IN GLASS. — SUCCESSFUL USE OF COLOR 329

CHAPTER XXX.

MACHINERY AND THE ARTS. — SCULPTURE WORK. — MOULDED BRICKS. — COLOR IN BRICK . . 335

THE ARCHITECT.

CHAPTER XXXI.

THE ARCHITECT AND THE HOUSEWIFE. — BUSINESS AND THE ARTS. COSTS. — PROPER UNDERSTANDING OF THE CLIENT'S WISHES. — PLENTY OF TIME TO MAKE PLANS 341

LIST OF ILLUSTRATIONS.

		PAGE
TAIL-PIECE	. . . . . .	16
FIG. 1. — BUILDINGS OF ALASKAN INDIANS	. . . .	18
FIG. 2. — A DECORATED ENTRANCE TO A CAVE	. . .	19
FIG. 3. — A GREEK TEMPLE AS AN ARTISTIC STRUCTURE	. .	19
FIG. 4. — A GREEK TEMPLE AS MERE BUILDING	. . .	20
FIG. 5. — A ROMAN DOORWAY	. . . . . .	21
FIG. 6. — AN OLD COLONIAL DOORWAY	. . . .	24
FIG. 7. — HOUSES AT CLUNY, TWELFTH CENTURY	. .	33
FIG. 8. — THE MUSICIAN'S HOUSE AT RHEIMS, THIRTEENTH CENTURY	.	34
FIG. 9. — FOURTEENTH-CENTURY HOUSE	. .	35
FIG. 10. — FLOOR PLAN	. . . . . .	35
FIG. 11. — HOUSE AT TREVES, FOURTEENTH CENTURY	. .	36
FIG. 12. — PLAN OF JACQUES COEUR'S HOUSE AT BOURGES	. .	37
FIG. 13. — BIRD'S-EYE VIEW OF JACQUES COEUR'S HOUSE, FIFTEENTH CENTURY	. . . . . .	38
FIG. 14. — AN EARLY FARM-HOUSE	. . . .	39
FIG. 15. — A TWELFTH-CENTURY HALF-TIMBER HOUSE	.	40
FIG. 16. — SECTION	. . . . . .	41
FIG. 17. — FRAMING DETAILS	. . . . .	42
FIG. 18. — A HALF-TIMBER HOUSE AT LISSIEUX, SIXTEENTH CENTURY	.	43
FIG. 19. — IN LISSIEUX	. . . . .	45
FIG. 20. — A STREET IN AURAY	. . . .	51
FIG. 21. — IN MALESTROIT	. . . .	53
FIG. 22. — A SHOP WINDOW IN JOSSELIN	. . . .	56
FIG. 23. — WINDOWS IN HÔTEL DES VOYAGEURS AT MORLAIX	. .	57
FIG. 24. — CHIMNEY-PIECE IN HOUSE OF ANNE OF BRITTANY AT MORLAIX	. . . .	58
FIG. 25. — A BRITTANY BED	. . . .	59
FIG. 26. — PLAN OF CHÂTEAU AND TOWN OF COURCY	. .	62
FIG. 27. — PLAN OF CHÂTEAU OF COURCY	. . .	63
FIG. 28. — BIRD'S-EYE VIEW OF COURCY	. . .	65
FIG. 29. — INNER COURT, COURCY	. . . .	66

1

		PAGE
Fig. 30. — Exterior Façade, Château of Josselin	.	67
Fig. 31. — Interior Façade, Château of Josselin	.	69
Fig. 32. — Bedroom, Château of Josselin	. .	72
Fig. 33. — Salon, Château of Josselin	.	75
Fig. 34. — Château of Pierrefonds		73
Fig. 35. — Château of Pierrefonds		76
Fig. 36. — Interior Façade, Château of Pierrefonds		77
Fig. 37. — Bedroom, Château of Chaumont . .	.	79
Fig. 38. — Bedroom, Château of Chaumont . .	.	83
Fig. 39. — Château of Chenonceau, River Façade .	.	85
Fig. 40. — Château of Azay-le-Rideau . . .		87
Tail-piece — A Bit of Decoration . .		89
Fig. 41. — Kentish Hall, Fifteenth Century . . .		93
Fig. 42. — A Seventeenth-Century House at Headcorn .		95
Fig. 43. — A Carved Spandrel		95
Fig. 44. — A House in Sussex		96
Fig. 45. — The White Lion — Crainbrook . .		97
Fig. 46. — House at Bittenden, Seventeenth Century		98
Fig. 47. — Ford's Hospital, Sixteenth Century .		99
Fig. 48. — Gable, Ford's Hospital		100
Fig. 49. — Bond's Hospital, Early Sixteenth Century		100
Fig. 50. — Gable, Bond's Hospital		101
Fig. 51. — Gable, Bond's Hospital . . .		101
Fig. 52. — A Coventry Gable		102
Fig. 53. — A Coventry Gable		102
Fig. 54. — A Modern English Interior . .		105
Tail-piece — A Decorative Motive .		107
Fig. 55. — Butchers' Guild House, Hildesheim		109
Fig. 56. — A Detail of Butchers' Guild House .		112
Fig. 57. — A Carved Doorway . . .		113
Fig. 58. — Decorated Panel		113
Fig. 59. — Carved Brackets .		114
Fig. 60. — House in Hildesheim		115
Fig. 61. — Overhanging Walls . .		118
Fig. 62. — A Corner		118
Fig. 63. — A Doorway of Carved Wood . .	.	119
Tail-piece	.	120
Fig. 64. — A Modern South German House in the Style of the Sixteenth Century . . .		121
Fig. 65. — House at Meiningen . .		125
Fig. 66. — House at Meiningen . .		126

Fig. 67. — Elevation		130
Fig. 68. — Floor Plan		130
Fig. 69. — Doorway		130
Fig. 70. — Doorway between Parlors		132
Fig. 71. — A Modern Example		133
Fig. 72. — A Brick House		135
Fig. 73. — First Story		136
Fig. 74. — Second Story		136
Fig. 75. — Looking from the Parlor		137
Fig. 76. — The Sitting-Room Mantel		138
Fig. 77. — Dressing-Room Closet		139
Fig. 78. — Closet with Doors Closed		139
Fig. 79. — Stair Window		142
Fig. 80. — Doors in this House		143
Fig. 81. — Doors in this House		143
Fig. 82. — First Story		144
Fig. 83. — Second Story		145
Fig. 84. — A Frame House		146
Fig. 85. — Looking from Parlor to Sitting-Room		147
Fig. 86. — Looking into Stair Hall		148
Fig. 87. — From Parlor to Stair Hall		149
Fig. 88. — Sideboard		150
Fig. 89. — Parlor Mantel		152
Fig. 90. — A Bedroom Mantel		152
Fig. 91. — A Moulded Band		153
Fig. 92. — Wall Covering		153
Fig. 93. — Sitting-Room Window		154
Tail-piece		155
Fig. 94. — Front Elevation of a Frame House		157
Fig. 95. — Side Elevation of a Frame House		158
Fig. 96. — A Dormer Window		159
Fig. 97. — Porch Details		160
Fig. 98. — First-Story String Course		161
Fig. 99. — Cornice		162
Fig. 100. — First Story		163
Fig. 101. — Second Story		163
Fig. 102. — Stairway		164
Fig. 103. — Hall Mantel		165
Fig. 104. — A Tile Mantel		165
Fig. 105. — A Tile Mantel	.	166
Fig. 106. — Door and Details	.	166

	PAGE
FIG. 107. — INSIDE DETAILS	167
TAIL-PIECE	168
FIG. 108. — FIRST-FLOOR PLAN	169
FIG. 109. — SECOND STORY	170
FIG. 110. — THIRD-FLOOR PLAN	170
FIG. 111. — EXTERIOR	171
FIG. 112. — LOCATION PLAN	172
FIG. 113. — HALL MANTEL	172
FIG. 114. — DINING-ROOM MANTEL	173
FIG. 115. — END OF LIBRARY	174
FIG. 116. — STAIRWAY	175
FIG. 117. — KITCHEN SINK AND TABLES	176
FIG. 118. — A DOOR	177
FIG. 119. — BARN, FROM REAR	178
FIG. 120. — BARN, FROM FRONT	179
FIG. 121. — BARN, FIRST STORY	180
FIG. 122. — BARN, SECOND STORY	181
FIG. 123. — A SCREEN	185
FIG. 124. — FIRST STORY	186
FIG. 125. — SECOND STORY	186
FIG. 126. — FRONT ELEVATION	187
FIG. 127. — FRONT ELEVATION	187
FIG. 128. — FRONT ELEVATION	188
FIG. 129. — FIRST STORY	188
FIG. 130. — SECOND STORY	188
FIG. 131. — STAIRWAY	189
FIG. 132. — PLAN OF STAIRWAY	189
FIG. 133. — FRONT ELEVATION	190
FIG. 134. — SIDE ELEVATION	191
FIG. 135. — FIRST STORY	192
FIG. 136. — SECOND STORY	192
FIG. 137. — LOOKING INTO ALCOVE FROM HALL	193
FIG. 138. — WINDOW-SEAT IN MUSIC-ROOM	194
FIG. 139. — SCREEN BETWEEN HALL AND MUSIC-ROOM	195
FIG. 140. — END OF DINING-ROOM	196
FIG. 141. — DOOR AND CASING	197
FIG. 142. — MANTEL	198
FIG. 143. — MANTEL	198
FIG. 144. — TAIL-PIECE	199
FIG. 145. — FIRST STORY	200
FIG. 146. — FRONT ELEVATION	201

	PAGE
FIG. 147. — SIDE ELEVATION	202
FIG. 148. — FIRST STORY	206
FIG. 149. — SECOND STORY	206
FIG. 150. — FRONT ELEVATION	207
FIG. 151. — FIRST STORY	208
FIG. 152. — SECOND STORY	208
FIG. 153. — FRONT ELEVATION	209
TAIL-PIECE — A BIT OF DECORATION	210
FIG. 154. — FLOOR PLAN	214
FIG. 155. — RANGE AND MANTEL	215
FIG. 156. — FIRST STORY	217
FIG. 157. — SECOND STORY	217
FIG. 158. — THIRD STORY	218
FIG. 159. — EXTERIOR	219
FIG. 160. — FIRST STORY	220
FIG. 161. — SECOND STORY	220
FIG. 162. — FIRST STORY	221
FIG. 163. — SECOND STORY	222
FIG. 164. — THIRD STORY	222
FIG. 165. — TWO DOUBLE HOUSES	223
FIG. 166. — PARLOR	224
FIG. 167. — SITTING-ROOM AND STAIR-HALL	225
FIG. 168. — A STORE BUILDING	227
FIG. 169. — LIVING-ROOMS OVER STORE	228
TAIL-PIECE — DECORATIVE MOTIVE	230
FIG. 170. — A PERSPECTIVE SKETCH	231
FIG. 171. — FIRST FLOOR	232
FIG. 172. — SECOND FLOOR	233
FIG. 173. — FRONT	234
FIG. 174. — SIDE	235
FIG. 175. — FIRST FLOOR	236
FIG. 176. — SECOND FLOOR	237
FIG. 177. — FIRST STORY	238
FIG. 178. — SECOND STORY	238
FIG. 179. — FRONT ELEVATION	239
FIG. 180. — FRONT ELEVATION	240
FIG. 181. — SIDE ELEVATION	241
FIG. 182. — OLD ROSES AND HOLLYHOCKS	243
FIG. 183. — OLD CONNECTICUT	244
FIG. 184. — ONE OF RICHARDSON'S HOUSES	245
FIG. 185. — A CITY PICTURE	246

			PAGE
Fig.	186.	— Studies	247
Fig.	187.	— A Seaside Picture	248
Fig.	188.	— In California	249
Fig.	189.	— Near Philadelphia	251
Fig.	190.	— A Thirteenth-Century Kitchen	254
Fig.	191.	— An Early French Fireplace	255
Fig.	192.	— Château of Courcy	256
Fig.	193.	— Fireplace, Château of Pierrefonds	257
Fig.	194.	— Fireplace, Château at Blois	259
Fig.	195.	— Fireplace in Cluny Museum	261
Fig.	196.	— Fireplace in Museum at Amsterdam	263
Fig.	197.	— Fireplace in Museum at Amsterdam	265
Fig.	198.	— A Hall Mantel	267
Fig.	199.	— In the Reception-Room	268
Fig.	200.	— Panels	273
Fig.	201.	— Panels	274
Fig.	202.	— Open Screen	275
Fig.	203.	— A Front Door	276
Fig.	204.	— Front Door	277
Fig.	205.	— Front Door	277
Fig.	206.	— Front Door	278
Fig.	207.	— A Brittany Door	278
Fig.	208.	— Decorated with Nail-Heads	279
Fig.	209.	— Doors	280
Fig.	210.	— Doors	281
Fig.	211.	— Doors	281
Fig.	212.	— Doors	281
Fig.	213.	— Doors	281
Fig.	214.	— Doors	281
Fig.	215.	— A Brittany Stairway	283
Fig.	216.	— The Landing	284
Fig.	217.	— At Nantes	285
Fig.	218.	— A German Stairway	287
Fig.	219.	— In the Museum at Amsterdam	289
Fig.	220.	— Book Shelves	294
Fig.	221.	— Brick Mantel	295
Fig.	222.	— Window-Seat and Bookcases	296
Fig.	223.	— A Cosey Seat	297
Fig.	224.	— A Seat at Pierrefonds	298
Fig.	225.	— A Couch	299
Fig.	226.	— A Recessed Window	300

Tail-piece — A Decorative Motive
Fig. 227. — A Twelfth-Century Grille
Fig. 228. — Of the Thirteenth Century
Fig. 229. — Common in the Twelfth Century
Fig. 230. — The Eleventh Century
Fig. 231. — Twelfth-Century Hinge
Fig. 232. — From Notre Dame, of Paris, Thirteenth Century
Fig. 233. — Of the Fourteenth Century
Fig. 234. — A Twelfth-Century Lock
Fig. 235. — A Thirteenth-Century Lock
Fig. 236. — A Fourteenth-Century Lock
Fig. 237. — Of the Fifteenth Century
Fig. 238. — Nail-Head
Fig. 239. — Nail-Head
Fig. 240. — Nail-Head
Fig. 241. — Nail-Head
Fig. 242. — Nail-Head
Fig. 243. — Nail-Head
Fig. 244. — Leaded Glass
Fig. 245. — Leaded Glass
Fig. 246. — Leaded Glass
Fig. 247. — Leaded Glass
Fig. 248. — Leaded Glass
Fig. 249. — Leaded Glass
Fig. 250. — Leaded Glass
Fig. 251. — A Sixteenth-Century Doorway
Tail-piece — A Decorative Motive

BEAUTIFUL HOUSES.

HOUSE-BUILDING AN ART.

CHAPTER I.

UGLY HOUSES. — UNEDUCATED ARCHITECTS. — THE ARTIST HOUSE–BUILDER. — THE MOST EXPENSIVE HOUSES NOT NECESSARILY THE MOST ARTISTIC. — THE ARTISTIC FAILURE OF GOVERNMENT BUILDINGS. — COST NEVER MEASURES THE ARTISTIC. — ORNAMENT AND LABOR. — ARTISTIC DETAILS COST NO MORE THAN UGLY ONES. — COMMERCIAL VALUE OF THE ARTISTIC.

MOST houses are ugly. One need only look at photographs or pictures of the principal streets of our cities and towns to see that this is true; the majority of the residences are commonplace, crude and pretentious. It is rare, indeed, that the most expensive house in a community is the most attractive. Good taste and large sums of money are rarely found together.

Our failure to make beautiful houses is certainly through neither lack of money nor lack of desire. Nearly every one who builds wishes to make an interesting structure. Few succeed.

We of America spend more money in building than any other nation, and the results are less satisfactory. We have only to consider the conditions in nations which do more artistic building to appreciate the reason for our lack of success. Our architects are not so well educated, nor are our people so appreciative. The community or the individual that builds must have a certain amount of artistic appreciation, and the architect must have an artistic training. These two conditions are necessary to good architecture. An uneducated public tolerates the uneducated architect, and the result is but a slow development in better things. The very large number of crude and uninteresting buildings degrades public taste. We are ambitious, but we are taught by common models.

One of good taste and fine feeling cannot hope to build an artistic house, excepting with the help of a trained artist. No one can lead his architect to do better than the architect knows. While the best results are reached through artistic sympathy on one hand, and artistic training on the other, it is possible for one quite insensible to beautiful things to build well through the dominating influence of a trained architect. However, a universal improvement in the character of our structures must come through the artistic advancement of both the public and the architect.

This leads to the statement that it is not the amount of money used in a building, but rather the taste displayed, which decides its artistic qualities. One without the artistic sense will generally do better work with a limited sum of money than with a more liberal allowance. With the smaller sum he is not able to express himself so conspicuously. A large expenditure of money enables one untrained in things artistic to display the real quality of his mind both through the size and the crudeness of detail of his building.

In any of the cities of our country the traveller will

always have the most expensive houses pointed out to him as the best. Such structures are generally as crude as they are costly. It often happens that such an one is surrounded by great lawns, beautiful drives, and attractive landscapes, but the house itself is a blot. In New York city the house built by W. H. Vanderbilt is very costly and pretentious. However, it does not rank high as an architectural production. There are scores of houses in that city which cost a mere trifle as compared with it, but which are highly successful.

This idea may be further illustrated by reference to well-known public structures. The Grand Central Depot in New York is expensive and pretentious. As an artistic production, it is not to be compared with the Boston & Providence Station in Boston, or the little stations built for the Boston & Albany Road by H. H. Richardson. The government buildings are generally inferior to the less expensive structures built by the State and city governments, or by private corporations. The national government has never built anything to compare with the City Hall in Albany, the Court House in Pittsburg, the Public Library in Boston, or the great office buildings of New York, Chicago, or other cities. The government building at the World's Fair was the least satisfactory of the more important structures. We may thus illustrate that with unlimited opportunity the result is unsatisfactory for want of artistic ability.

The same principle applies to all buildings. It has its relation to low-cost houses as compared to those of greater expense. We can never measure the artistic value by the cost. Of two cottages, one costing twenty-five hundred and the other six thousand dollars, the latter may be a failure and the former a success. It is more a question as to the way in which the material and labor representing certain values are used than its quantity. Money gives many people an opportunity to display bad taste.

There is a firm impression in the minds of the public that a house, in order to be attractive, must have a certain amount of what is called ornamental work applied to it. This is generally separate and apart from the structural parts. The house is built and the embellishment merely added. These details are not ornamental excepting in name. The plain surfaces are merely disturbed. They are cut into unusual forms without the least display of artistic sense. The different parts are expressive of complexity, but rarely of beauty. All of these things call for relatively large expenditures. There having been no artistic guide, the work is only laborious. Beauty and attractiveness are entirely wanting.

A careful study of this subject will develop the fact that ugly houses are built through a misdirected effort to make them pretty rather than through indifference. The lack of success is the result of untrained effort. Houses are more frequently ruined by spending too much money than they are through not using enough. Under proper guidance these expenditures might have been to enrich and beautify them. As it is, they have only made them cruder. Any one can see houses, large and small, which would be much better from an artistic standpoint if the owner had used less money.

In this is not to be found an argument against the decoration of buildings. It is an argument against adding ugly details to plain surfaces. It is intended to suggest that the mere sawing of a board into strange and unusual shapes adds to the cost of a house at the same time that it detracts from its value.

This is best understood when we look at the house in detail. Nearly all cornices represent a mere expenditure of material and labor. The finish of doors and windows is simply complicated. As to gables, the effort is largely directed to making them unusual. Porches are combinations of crude parts. In truth, in most houses the labor is entirely without artistic direction. The artistic builder — the architect — looks first to a pleasing outline, and instinctively selects that which is relatively simple. He so adjusts the constructive forms as to give pleasing details. Where there is the opportunity for decoration, the forms are beautiful in themselves and carefully studied in their relation to other parts. Having developed his outline, the different details are considered in their proper relation. In decorating a gable he preserves its natural forms and outlines. If he chooses to decorate the rafters, he does it in the same careful, artistic spirit that a painter would execute a very important part of a picture. The brackets, the mouldings, the windows, all show the mind and the heart of the artist. This same high spirit pervades everything — the doors, the windows, porches, balconies. Each part is considered in its relation to the whole and the composition. This is the way the artist works. Abundant means may enrich the detail, or may affect the character of the outline, but the artist shows his character under any circumstances; there are no limitations of cost which will entirely suppress this spirit.

It costs no more money to make a pretty, graceful moulding than it does to make a clumsy and awkward one. The same machine and the same power will give an attractive form as readily as an ugly one. It is simply a question of training and direction of mind. It costs no more to cut a pretty piece of scroll work than one in which there is no thought of beauty, but there must be the trained hand to direct it, to draw the design. Pleasing forms can be carved at the same cost as ugly ones.

These are matters of detail. We can consider the different parts of a house and affirm the same truth with respect to each particular part. It is plain that if in detail it costs no more to make what is pretty than what is ugly, in the aggregate it costs no more to make a pretty house than an ugly one. It is the mission of this book to make this fact plain.

Good design adds to value. The artist directs the labor and gives pleasing forms to the material. One not trained directs labor and gives to material forms not pleasing. The former through the same expenditure of material and labor produces what is more valuable than the latter. In every city there are a large number of buildings which are worth very much less than they cost. The design is bad and the construction faulty. These structures represent a positive waste. The artist did not direct the work; hence a loss. One may look around him and see two houses which cost the same sum of money, say four thousand dollars One is a mere building and the other an artistic production Each contains material and labor which cost four thousand dollars. One is built by an artist and the other by a builder. The former is worth much more than it cost and the latter much less.

We sometimes hear houses spoken of as being out of

style. That cannot be. Good art, good construction, and good floor planning never go out of style. The history of architecture affirms this. Many buildings have been constructed in past centuries upon which we look now as ultimate achievements. The commercial value of the artistic is not fully appreciated.

As a matter of business one who, with the help of an artist, spends eighteen hundred dollars in a house has a much better structure, when viewed from a money standpoint, than one who merely gets eighteen hundred dollars of crudely applied material and labor in house form. In how much better position are those who employ well-trained architects to do their work than those who operate without such help. The design and arrangement of a house are of more commercial importance than the material and labor which form it. It is the thought which gives value.

The ideas of design which have been set forth in this connection are as old as art, but the public has been misled in the matter. The spirit of extravagance has associated the ideas of beauty and great cost. They have been considered inseparable, and in the struggle for luxurious effects with limited means, we have a tawdriness and commonness pervading the majority of houses built. This, combined with the general ignorant struggle, has, through sham effects, brought our domestic architecture into a rather low state. Many who are known as architects have helped foster this spirit. Relatively few are educated in the true artistic sense, and the majority progress in the same spirit of constructive crudeness which is common to mere builders. We have only to take up the architectural journals of to-day and look at the designs of houses published to realize the true state of affairs. It indicates that the majority of architects are pampering the public demand for fussiness and sham luxury. It is to an improved public taste and a few conscientious architects that we may look for reform in this matter.

General culture or education does not bring with it artistic knowledge, appreciation, or judgment. This may be illustrated by an examination of the principal buildings of our great educational institutions. The very early buildings of Harvard College, for instance, are among the best. The general statements made will apply to most educational institutions. Their architecture is usually bad, proving as conclusively as possible that general education does not correspondingly develop an artistic sentiment. Artistic knowledge comes from a direct association with things artistic.

HOUSE-BUILDING AN ART. — *Continued.*

CHAPTER II.

DEVELOPMENT OF ART IN BUILDING. — THE PRIMITIVE HOUSE. — FIRST PRINCIPLES. — THE GREEK TEMPLE AND THE INDIAN HUT. — THE FUNCTIONAL AND THE ARTISTIC. — ROMAN METHODS OF BUILDING. — ACADEMIC ARCHITECTURE. — THE OLD ROMAN AND THE OLD COLONIAL. — ROMANESQUE ARCHITECTURE. — GOTHIC ARCHITECTURE. — EMOTIONAL EXPRESSION IN BUILDING. — DECLINE OF THE GOTHIC. — THE RENAISSANCE. — MODERN ARCHITECTURE OF EUROPE. — CHARACTERISTICS OF MODERN AMERICAN ARCHITECTURE. — AMERICAN STRUCTURAL FORMS. — AMERICAN METHODS OF LIVING AS RELATED TO ARCHITECTURE. — RELATION OF PRECEDENT TO FUTURE DEVELOPMENT.

ONE reason why our buildings have been so generally inartistic is that people, as a class, do not realize the relation which exists between art and building. We cannot expect to reach this result by making working artists of the people. It is not necessary that one should be able to draw or to paint in order to give the mind an artistic training. One may be an artist in heart and mind without manual dexterity in the arts. It is through artistic cultivation that we may expect an improvement in architecture. Good taste is not inherent; it comes by direct association with things artistic. The intelligent sympathy and support which the people may give their architects and builders will do quite as much to remove the present unsatisfactory condition as anything which can be thought of at this time. It is to give a general idea of the development of architecture and the relation of art to building that this chapter is written. The primitive houses, a picture of which is given in Fig. 1, were constructed by Alaskan Indians. They might as well have been the structures of a thousand or two thousand years before Christ. It is a primitive idea. It illustrates the first principle in building. The walls are made of any

FIG. 1.—BUILDINGS OF ALASKAN INDIANS.

convenient material. The porch is constructed with tree trunks for columns and branches for lintels.

Another primitive dwelling is a cave. The ceiling forms a natural arch; a few stones are jambed together at the doorway, and we have an arched opening. In these two kinds of houses we have all of the principles of building: columns for support, and lintels which rest on them; or an arch over a doorway, which carries the weight above. These are the only known principles of construction. The rest is simply detail. The principle of construction involved in the hut of the Alaskan Indian is the only one used in the construction of the principal temples of Egypt and Greece. The arch with the lintel was used in the Roman structures, the cathedrals of the Middle Ages, and those of the Renaissance. The principles of the lintel and the arch are practically the sum and substance of building. Fig. 2 shows the decoration of the entrance to the cave. When the work of the artist and that of the builder are united, we have architecture.

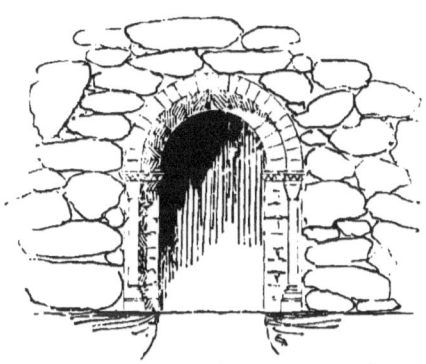

FIG. 2.—A DECORATED ENTRANCE TO A CAVE.

The front of the Greek temple, Fig. 3, does not look much like the Indian hut, yet no one can deny the identity of the principle of construction as shown in porches. Fig. 4 shows a Greek temple as it would appear as a mere building with the art omitted. By looking at it we see how coarse and common a Greek temple might have

FIG. 3.—A GREEK TEMPLE AS AN ARTISTIC STRUCTURE.

been. The great columns of marble, the ponderous lintels, the heavy cornice. and the expanse of roof are capable of being made very common through the treatment of the detail. However, the Greek took this structural form and added to it more of delicacy, nice discrimination, and gentleness of expression than the world has since known in art. Theirs was an exclusively intellectual art with an absence of emotion. There was the splendid, robust construction, the immense columns of marble with great lintels extending from one to another, and above all a mass of pediment beautifully decorated. It was a fine and delicate mind which took this rugged outline and refined every detail. This made Greek architecture.

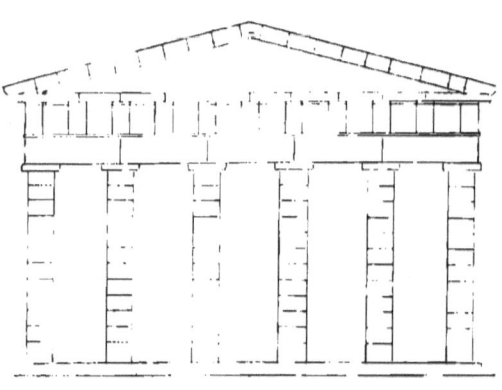

FIG. 4.—A GREEK TEMPLE AS MERE BUILDING.

I have mentioned enough of the principles of good architecture to give an understanding of what is needed in our American homes. A house should be arranged to meet the requirements of its occupants. So was the Greek temple. It should be constructed in a direct, substantial manner. So was the Greek temple. It should be decorated in a thoughtful, refined spirit. So was the Greek temple.

The Romans took the Greek orders, and in the course of a few hundred years surrounded their reproductions with about the same kind of order and system that we find in a

FIG. 5.—A ROMAN DOORWAY.

well-organized manufactory of our own time. By this expression I do not wish to belittle the character of the Roman work. It was monumental, grand, impressive, and artistic, though in an essentially different way from the work of the Greeks. They had the help, originally, of the Greek artists; but the organizing quality of the Romans led them to formulate everything. It obliterated the Greek method and gradually developed an architecture which was entirely dependent upon formula. Their work was scholastic. To be a Roman architect it was not necessary to be an artist at heart. It was a question of knowledge rather than of feeling or impression. They were great engineers with a splendid academic knowledge of architectural forms. Theirs was an academic architecture to which any one could be trained. Their drawings were prepared by educated mechanics who worked according to precedent and rule. The work was executed by other mechanics.

This general system could be applied to the building of American houses. House-building knowledge could be formulated, worked into text-books, the decoration thereof applied according to method, and we should have an architecture more precise, infinitely more satisfactory than the hodge-podge which stares us in the face every time we go on the street. The Old Colonial architecture originated in a knowledge of classic forms. It was an application of academic architecture to the decoration of American homes.

Fig. 5 is an old Roman doorway; Fig. 6 is taken from an old colonial house in New England. One is constructed in stone, the other in wood. The similarity in form is apparent. When we speak of this architecture as academic, we mean that it can be constructed according to rule.

There are definitely expressed directions for the relative size of the columns, the proportion of the cornice to the height of the columns, and the exact method to be used in the production of each particular moulding and other decorative feature. This is a formal architecture — not to be commended as the best method, but merely mentioned as one method, and from which our Old Colonial style developed.

If the record of the world's best productions and the methods of their development were brought together there would be developed a text-book architecture which would be far from offensive, but which would not be spirited, and certainly not American in character.

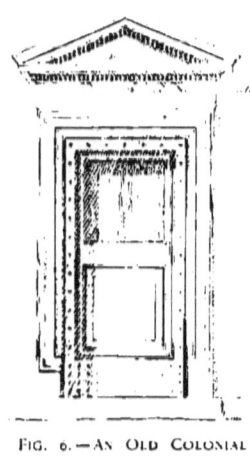

FIG. 6.—AN OLD COLONIAL DOORWAY.

The Roman architecture expired with the decline of that empire, and it was not until the eleventh century that there was general evidence of architectural development. The architecture of the eleventh and twelfth centuries is known as the Romanesque. We must not, however, too closely connect this Romanesque architecture with that of the Romans. Whilst it had some of the Roman forms of decoration, there was more of freshness, vitality, and originality in the Romanesque architecture of the eleventh and twelfth centuries than belonged to that with which its name is so closely allied, but from which it was so widely separated.

The Romanesque decorative forms in a refined form were made known to us by our great architect, Mr. H. H.

Richardson. The Trinity Church of Boston, the City Hall in Albany, the Chamber of Commerce at Cincinnati, the Court House at Pittsburg, Marshall Field's store in Chicago, and many other artistic structures were designed by Mr. Richardson, who had many small followers and imitators throughout the country.

In the thirteenth century we find a great climax. This architecture had more of good construction, originality of form, and artistic decoration than any we have since seen. It is an architecture overflowing with ideas. It was during that period that the great Cathedral of Notre Dame at Paris, those at Bourges, Rouen, Amiens, Rheims, and hundreds of others were constructed. It came as a blaze of light out of the Dark Ages. The Romanesque was a mere glimmering of what was to follow. The Gothic of the thirteenth century was the grandest illumination the world has ever known. It is startling from its originality. Constructively, it was radical in all its details. Its decoration was effervescing with newness and beauty. It was an architecture born of a fervent religious sentiment; the people poured themselves into it. It is the expression of high emotion in building. Many of our churches have followed the style of the Gothic of the thirteenth, fourteenth, and fifteenth centuries, but they pale into insignificance when compared with the architecture of that period. The splendid spirit has ever since been lacking. There is to-day more of art in a thirteenth-century cathedral than is to be found elsewhere. Where else does one find architecture, sculpture, painting, music, and the splendid ceremonial in so high a state of development? One cannot hope to find elsewhere anything which will have so great an effect upon the emotions.

In the fourteenth and fifteenth centuries the Gothic architecture became too luxuriant, and in the sixteenth it gave place to the Renaissance. This early Renaissance architecture borrowed its decorative forms from the old Roman period; yet in France it embodied the spirit and vivacity of the Gothic. It was all borrowed, but the borrowers displayed wonderfully good taste; they made it more beautiful in detail, more original in general conception, than was the source of their general idea. The spirit was extinguished in the seventeenth century, and in the eighteenth century it was dead. Under the guidance of McKim, Mead, & White, of New York, and others, there has been developed during recent years a sincere, studious revival of the architecture of the Renaissance. Most of the work of the Chicago Exposition was in that style. This illustration may serve to bring it measurably to our minds.

I have indicated in a broad way the lines along which architecture has been developed. I might have spoken of the Byzantine architecture — an art having Roman structural forms and much of the fineness and sensitiveness of the Greek — or of the luxurious architecture of Spain; but there is no necessity for this. Historical architecture, as we generally know it, must be either Greek, Roman, Romanesque, or Renaissance.

We read of the style of Francis I., of that of Louis XIII., XIV., and XV., of the Dutch Renaissance, of the Queen Anne and Elizabethan; but all these belong to the Renaissance. They are mere parts of the development and decline of that architecture. The architecture of the early part of the Renaissance is much better than anything which developed later, as was the architecture of the early part of the Gothic period superior to what followed. It is natural for us to ask about modern architecture in the Old World. With all their splendid examples, out of the research of the past, what are they doing to-day? To speak plainly and directly, we must say that the modern architecture of Europe is relatively inferior. We find no great structures in the style of the Renaissance which equal those of the early Renaissance. We find no structures in the Gothic style equal to those of the thirteenth century. In the art of architecture we have to look backward several hundred years to find higher development. Viollet-le-Duc has said: "In architecture alone man looks back upon the masterpieces of the past, not as points of departure, but as ultimate attainments, content for his own part if, by recombining the elements and reproducing the forms of these monuments, he can win from an esoteric circle of archæologists the praise of reproducing some reflex of their impressiveness."

If we speak of the architecture of the immediate present, in Paris, for instance, we say at once and unhesitatingly that it is not ugly. The details are usually good, the carving and decorative work never clumsy, and none of the finer details exaggerated or unusual. The School of Fine Arts, in Paris, has left an indelible imprint upon the modern architecture of France. This influence is not to be exaggerated. The school architecture is the architecture of that country. To say that the modern commercial architecture of Paris is not monotonous would be untrue; to say that it is ugly or clumsy, that its details are bad or conceived in ignorance, would be equally untrue. There is the monotony of uniformity, but never ugliness. There are miles of seven and eight story structures with no effort towards novelty, no attempt at wildly picturesque effects, but with never an offence against refinement of detail. The mouldings, the carving, all of the decorative features, as well as the composition, are exact, studious, refined, and usually beautiful.

A great many people say that the modern work of Paris is monotonous, but no one ever says that it is ugly. It is the product of a single great school which has only one idea — to teach classic architecture. There is a soberness, a quietness, and a dignity about its products which, when they are numerously multiplied, give little variety or picturesqueness. This is measurably true of all the architecture of the Continent. It is all dominated by the great school of France. While this is not so true of the architecture of the United Kingdom, it is a fact that what this section gains in picturesqueness, through the lack of a great sober impulse, it loses in refinement. The city architecture of Great Britain is only narrowly removed from the uncertainty, the lack of repose, and the crudeness which belong to so much of the architecture of America.

The earlier architecture of this country is known as the Old Colonial. It is the work of highly educated carpenters and mechanics, who came to this country during the Old Colonial period. The style is of classic origin, the details belonging to the Roman and Greek architecture. Generally speaking, they were admirably simplified, and served a splendid purpose in the decoration of our American homes. There is a quietness and sobriety about this architecture, a stamp of knowledge, which naturally brings it into great favor with our people. It was the product of academic builders, those well schooled in their work, and it is the best architecture we have ever had for our dwellings.

A great many people will ask about an American style of architecture. In a measure we have it, not generally, however, in a state of great refinement. The idea is new, original, and promising. Our dwellings, such as are now designed by a small number of our best architects, are artistic productions. They are entirely distinct from anything known to the rest of the world. But this is only a part of our architecture. Our steel and burnt-clay construction belong to America. The commercial structures of our great cities are unique as commercial productions. There is erected, first, a self-supporting steel frame, which is surrounded by some form of burnt clay, brick, terracotta, or stone. A style of architecture is developed from a nation's distinctive requirements. Our commercial demands present a new architectural problem. We have met it successfully. It is true that the clothing of these steel structures is often rather crude, but this is not universally true. Given the steel frame and an artist to decorate it with structural coverings, and we have a well-established style of American architecture. In our Protestant churches we are departing from the general plan suggested by the old cathedrals. The long rooms and tall spires have no place in meeting the requirements of a Protestant congregation to-day. The auditorium, constructed solely with a view to seeing and hearing, of a character to seat the greatest number with greatest comfort, is a necessity of our own times. The artist who takes the forms suggested by these requirements, and properly decorates them, is doing his best to develop an American school of architecture. But, after all, an American school is not important. It is good architecture that we want, from whatever source.

Primarily we must meet the physical requirements of our structures in a direct manner, and then decorate them with the best motives which the world's architecture has to offer us. If we can do this in an original spirit, it is well, but originality is not essential.

There is an inclination upon the part of many of us to look for the development of American architecture along the lines which architecture has taken in the past. There is no good reason for this. The cathedral, the château, and the great historic monuments of the Old World were constructed in response to conditions which can never come to us. Our methods of living, our commercial conditions, the tendency of religious thought, our educational system, and, in truth, our general social condition furnish problems for the architect varying greatly from those of the earlier periods. The character of the architecture will be quite as individual as the varying conditions which give it birth. Instead of great military châteaux, we have thousands of towns and cities constructed without regard to defence, and an architecture which leaves the military conditions entirely out of consideration. It is an architecture of peace.

Instead of the great cathedral, with its art, forming a part of the impressive religious service, and having a splendid emotional effect upon the people, we have the more modest chapels, lecture halls, and auditoriums wherein people are instructed without serious regard to the emotional influences. This is a condition which makes a great change in the architecture. Our rapid means of communication from one section to another, the trade conditions, the peaceful relations with the people of the various sections, the more uniform distribution of wealth, and the education of the masses have developed a commercial architecture which has never before had a proper place in the world's history. Our educational methods create the necessity for the construction of many buildings of a character not needed in earlier times.

Thus it is readily seen that we are developing structural necessities which, if handled in a perfectly direct, unaffected, artistic manner, will produce an architecture having its own character and belonging distinctively and solely to our own times. It is our clinging to the past which binds us. It is our blind study of precedent which keeps us stationary. Precedent has distinct value, but only when intelligently used. A French château is not suited to the conditions of our living at this time. However, many of the decorative details of these buildings will help us clothe structures arranged with reference to our own demands. The Old Colonial architecture is the most artistic expression of the home condition, and home life of a people, which we have here. The houses were planned with reference to the uses intended, and the beautiful classic details were so modified as to decorate them, express the domesticity, home comfort, general refinement, and splendid character of the people who lived at that time. No architecture can be more artistic, none can be more highly commended, than one thus successful in its expression.

In this I have undertaken to make plain the sources from which we can draw our architectural inspiration, the lines of the development of the architecture of our own times, and the definite relation which present demands bear to architectural forms.

THE WORLD'S HOMES.

CHAPTER III.

FRENCH DOMESTIC ARCHITECTURE. — NATIONAL BUILDING METHODS. — TWELFTH-CENTURY BUILDING. — THE MUSICIAN'S HOUSE AT RHEIMS. — FLOOR PLANS OF DOMESTIC STRUCTURES. — PICTURESQUE STAIR TOWERS. — THE HOUSE OF JACQUES COEUR. — AN EARLY FARM-HOUSE. — HALF-TIMBER ARCHITECTURE OF THE TWELFTH CENTURY. — LATER DEVELOPMENTS IN WOODEN ARCHITECTURE. — OUR USE OF FRENCH EXAMPLES. — FROM LISSIEUX.

FRANCE presents a wealth of material which may be studied in connection with domestic architecture for its unequalled variety and its artistic qualities. Excepting the châteaux, the structures of which we have the completest record are almost entirely buildings fronting directly on the street or road. In France it is rare indeed that one sees an isolated building with a free passage around it, as is common in our American towns and cities. It is not at all uncommon for a farm building to be constructed within a wall; again, the farmer's house may be almost flush with the road. Little farm communities, with the buildings abutting on one another, are very common, because of the companionship which such association brings. This was not alone true in the early history of France, but obtains in the constructions of to-day. The small towns, as well as the cities, are almost universally built very compactly. Thus we may expect to see very few examples of isolated structures in France. There are, however, a few field-houses of which I give a typical illustration, which are interesting as showing the early methods of building such houses.

FIG. 7 — HOUSES AT CLUNY. TWELFTH CENTURY.
(From " Architecture Civile et Domestique," Verdier et Cattois.)

The centre of development of civil, domestic, and religious architecture was Cluny. There is a great temptation to select a number of illustrations from this city. However, one must suffice — a stone building, two stories in height, and covered with a tile roof. The front of the first floor is used as a store ; the back part, no doubt, is a living-room ; the second floor is manifestly used for domestic purposes (Fig. 7). We have here the spirited composition and rugged vigor which were a part of the Romanesque

FIG. 8 — THE MUSICIAN'S HOUSE AT RHEIMS, THIRTEENTH CENTURY.
(From "Architecture Civile et Domestique," Verdier et Cattois.)

architecture ; a style having its origin in the eleventh century and extending through the twelfth. It had its own natural constructive form, but borrowed many of its decorative details from the old Roman architecture. Certainly the general form, the mouldings, and often very many of the decorative details were imported from the Roman styles. However, in this connection it is well to bear in mind that the general constructive details were not borrowed, but had their origin in the practical necessities of the time in which they were devised.

The Musician's House at Rheims (Fig. 8), which belongs to the thirteenth century, is typical, in the grace of its detail, in its beauty of composition, and the high character of its sculpture, of the architecture of that time. It is a particularly valuable example, because it admits of adaptation to our own times.

From the architecture of a hundred years later we select a building of most ingenious plan. This will be noticed in the arrangement of the stairway and entrance (Fig. 9).

FIG. 9 — FOURTEENTH CENTURY HOUSE.
(From Viollet-le-Duc's "Dict. de l'Architecture.")

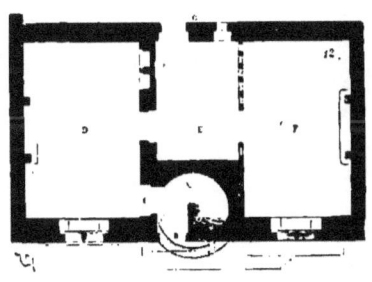

FIG. 10.

This is another of those very useful examples which may be readily adapted to our own times. One cannot but call attention to the simplicity of the general composition; the arrangement of parts, and the beauty and gracefulness of the various details; the mouldings and finish around the windows; the finial of the little tower, the chimney tops and the gables. The string courses are fine in character, the hinges on the front door particularly graceful, and the finish on the tower above it one of those startlingly practical and original devices which cannot but command respect when carried out in this perfectly natural, straightforward manner. With the exception of the entrance to the cellar, all of the windows have square openings. All are quite simple in character and relatively free from great expense in construction (Fig. 10).

FIG. 11.—HOUSE AT TREVES, FOURTEENTH CENTURY.
(From Viollet-le-Duc's "Dict. de l'Architecture.")

I cannot refrain from selecting another example of this century — a three-story stone building from Treves. It is a most charming and graceful structure, which must commend itself to all people of good taste. Even in the small cut we see the delicate character of its detail, the beautiful spacing of its openings, and the recognition of all constructive necessities. It is stamped throughout with the honor and refinement of a true artist; and withal there is its simplicity and availability for our own needs. There is no better opportunity of again emphasizing the fact that here is a relatively inexpensive structure which is a great artistic success, and one upon which, if the artist had had much larger material resources, he would not have found it advantageous to have spent an extra cent. It is simple, beautiful, inexpensive, and entirely successful (Fig. 11).

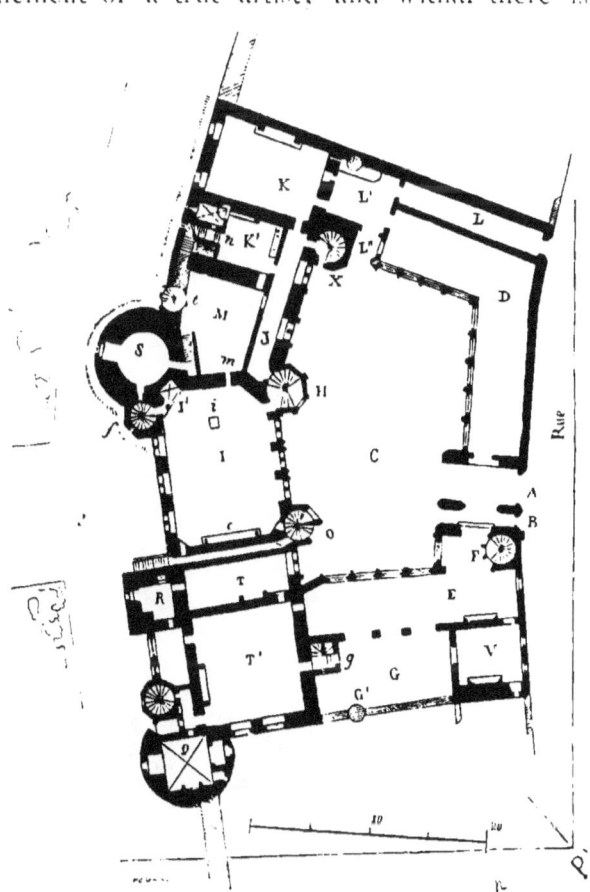

Fig. 12.—Plan of Jacques Cœur's House at Bourges.
(From Viollet-le-Duc's "Dict. de l'Architecture.")

One of the most magnificent private establishments of the fifteenth century is the house of Jacques Cœur of Bourges. He secured the right to enclose a part of the ramparts of the city as one wall of his building. The towers Q, R, and S were the original towers of the old wall. The general plan of this structure (Fig. 12) is typical, in a broad sense, of the arrangement of the ambitious private establishments of that time. The principal entrance is shown at A. The court is C. There are colonnades surrounding two sides of it, and rooms connected with it on all sides. Stairways to the various apartments are shown. Practically each separate apartment has its own stairway.

FIG. 13.—BIRD'S-EYE VIEW OF JACQUES COEUR'S HOUSE, FIFTEENTH CENTURY.
(From Viollet-le-Duc's "Dict. de l'Architecture.")

The bird's-eye view (Fig. 13) gives a very fair idea of the outline of this building, which, by the way, is in a state of fine preservation to-day. The decoration is extremely rich, and the outlines are ideally graceful. One would search in vain for an awkward moulding, an unfortunate bit of decoration, or any other departure from the work of a great artist.

Fig. 14 is a typical illustration of one of the early field or farm houses mentioned on page 32. The artistic instinct is apparent even in this structure. The chance for display of the picturesque is always accepted, and in this simple building

FIG. 14.—AN EARLY FARM-HOUSE.
(From Viollet-le-Duc's "Dict. de l'Architecture.")

are many interesting suggestions, and certainly a splendid example of what may be done by very simple means.

Half-timber architecture was common in France until something over a hundred years ago. The general character of its construction has always been the same, though in different periods it has varied somewhat in detail. The earliest example of this kind of work is cited by Viollet-le-Duc, and had its origin in the middle of the twelfth century (Fig. 15). He sketched it in 1834, at which time it was being taken down to give place to another structure. It is to be noticed that it is a three-story building, with the two upper stories overhanging. The partition walls on each side are of stone, and the projection of the wall is carried on corbels. The section and details show the general character of the construction. Fig. 16 is the section, and indicates in outline the character of the framing. On the first story, as shown in the section, A is a post, B a bracket, R a girder, D a cross-sill running over the girder. Above the girder are intermediate supports. The support P is mortised and pinned to the girder L, and rests on the piece on girder at R. M is the notch cut in the upright E which carries the girder L. The piece E continues through to the sill F, and the bracket I rests on a corbel O. The other details of construction on this section will be understood by merely looking. The square pieces corresponding to our joists, which rest on the sills, reach from the intermediate girder to the walls on each side. Fig. 17 shows the method of framing between the windows, the piece immediately above

FIG. 15.—A TWELFTH-CENTURY HALF-TIMBER HOUSE.
From Viollet-le-Duc's "Dict. de l'Architecture.")

them and the cross-girder which is above the window-head. The tenons I and B reach above the top of the window-head A, and in so far as they project above this piece they are inserted into the mortise E. The space on each side of the window, which is shown at E, is let into the upper part of the post G. The tenon M is inserted into the mortise. The thoroughness of this construction cannot but command the respect of any observer.

The cuts are clear enough, so that if any one who is sufficiently interested will look at them for a few minutes, they will be readily understood. Here is a building constructed in the twelfth century — a wooden building sketched by Viollet-le-Duc in the thirty-fourth year of the nineteenth century. It should put to shame our modern idea of wooden construction. We have buildings erected in the eighteenth century, the Old Colonial construction, which in a few years will have fallen into decay. In France wooden constructions are in a good state of preservation which had their origin in the fourteenth, fifteenth, and sixteenth centuries. They are now occupied and have rental value.

It is not a question as to the selection of examples: it is one of discrimination. When we present buildings representative of the domestic architecture of Brittany, of middle and northern France, of Germany, Switzerland, and England, it is not with the idea of showing variety or the cumulative information of these various sections.

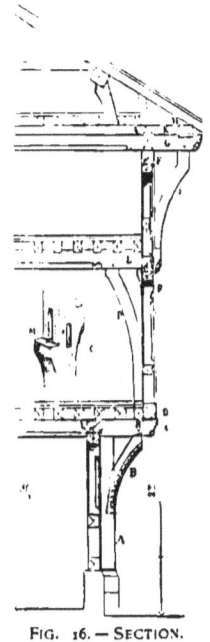

FIG. 16. — SECTION.

Quite the contrary. It is to show what these various countries have done, with independent knowledge; how they have operated from their varying standpoints, and in their independence, not knowing what has been done elsewhere, have been successful. They did not have the literature of their architecture crowding in upon them from many sources every week. Yet without such facilities they developed an architecture much more successful, altogether more satisfactory, than that which has been developed under the conditions of the present time.

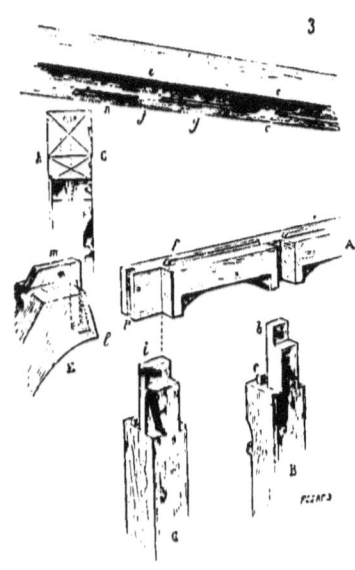

FIG. 17.—FRAMING DETAILS

Fig. 18 shows a half-timber structure at Lissieux, a very picturesque city of northern France. It was built during the sixteenth century.

One might fill many pages with the beautiful details of this structure, but the general expression of the exterior given in this illustration must suffice. The framework is of wood, and the filling between of brick. The carving is of a most exquisite character. All of the mouldings are from the hand of refined, sensitive, appreciative artists.

In our own time we have many picturesque shingle structures. When they are developed naturally by an artist, they are our most successful domestic structures. However, it is not a question of shingles or cement, brick

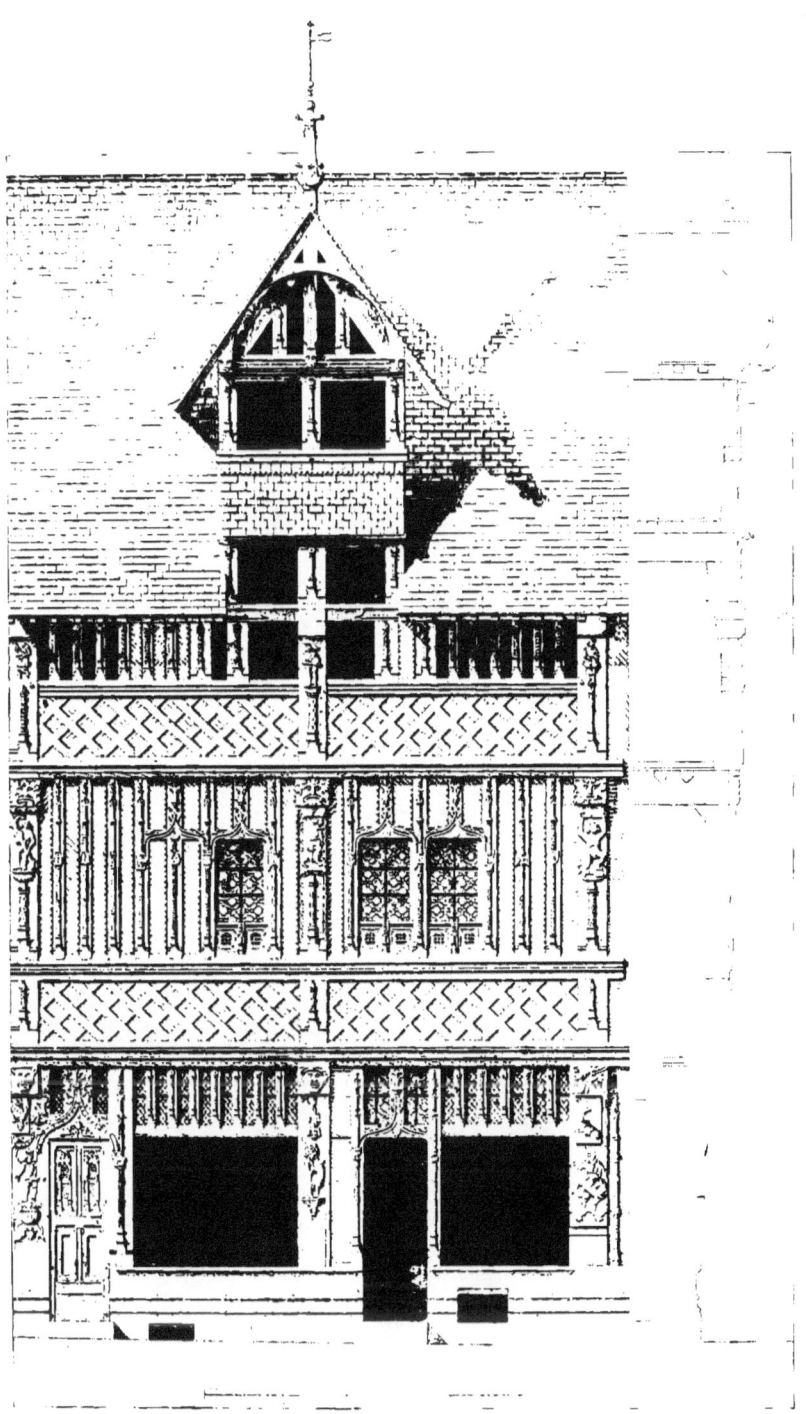

FIG. 18.—A HALF-TIMBER HOUSE AT LISSIEUX, SIXTEENTH CENTURY.
(From "Architecture Civile et Domestique," Verdier et Cattois.)

FIG. 19. — IN LISIEUX

or other material. It is one of the quality of mind which directs their use.

In Fig. 19 is given another example from Lissieux. In it is a half-timber structure, with the various details executed in slate. Other portions between the structural timbers are of masonry and plastering. In the left-hand corner is a very charming example of ironwork of the seventeenth century. The other work referred to belongs to the sixteenth century. It must be borne in mind that the selections given do not pretend to be of the best or the richest of the various types, but are rather typical examples selected from a collection so large that there is great difficulty in actually presenting the most satisfactory.

The history of half-timber architecture alone might be made to cover many volumes larger than this. These first chapters are intended to be illustrative, to show how the various sections of the world, not widely separated by distance, though distinctly separated as to means of communication, without the literature of architecture which belongs to our own time, were enabled to develop work of a character more artistic, more beautiful, occupying a larger place in the world's history, than we can expect to find in the present century.

We have cause for humiliation when we see what has been done in domestic architecture in France, England, Germany, Holland, and in our own country during the Old Colonial period, and compare it with the work of our own times.

THE WORLD'S HOMES. — *Continued.*

CHAPTER IV.

BRITTANY. — THE PAINTER'S COUNTRY.—FEW OFFENSIVE BUILDINGS. — CHARACTERISTICS OF THE PEOPLE. — THE ARCHITECTURE, THE SCENERY, AND THE PEOPLE HARMONIOUS. — BRETON COSTUMES. — FURNITURE. — EXTRAVAGANTLY PICTURESQUE. — MALESTROIT. — A CHIMNEY-PIECE AT MORLAIX. — BEDS.

THE student has not been directed to that part of western France known as Brittany to study the picturesque in architecture. Painters from many parts of the world go there, and Brittany is well known to us through their canvases. Jules Breton, one of the great poet-painters of our own times, saw the splendid character of the Breton peasant so clearly that much of his best work came from that section. Our own Walter Gay has presented, through the medium of his brush, great thoughts which were inspired by Brittany. In that beautiful country are many American art-students, who find much in it to inspire them. The world knows its beautiful scenery and its splendid people through their efforts. The architecture, however, has not played a great part on their canvases. Yet in all Europe there is nothing so novel and so quaint and odd as this Brittany architecture. Withal, there is much of it that is very beautiful. One sees many towns and cities in Brittany in which there is not one offensive structure. All are interesting, the large portion very picturesque, and many exceedingly beautiful. A glance at the illustrations used in this chapter will justify all that I have said with respect to this architecture.

Yet to understand it better one should know something of the people. That part of the world is to-day three or four hundred years behind ours in commercial movements and habits of living. There one sees men and women weaving by hand, threshing grain with flails, cutting wheat with a sickle, sawing lumber by hand. The manners of the people, their dress and their architecture, belong to the same period — the fourteenth and fifteenth centuries. They are measurably influenced by legends which date from the ninth and tenth centuries. The costumes of the men and women are always picturesque and nearly always beautiful. The change in their costume is by geography and not by time. In riding or walking over twenty miles of country one will pass through four or five sections clearly distinguished, one from another, by the costumes of both men and women. The dress in each particular section is the same year after year, yet it varies with every small section. The picturesqueness is not alone in the form of the various garments, but as well in their color. They often abound in splendid embroideries and other decorative features. The design of these costumes is usually very pronounced, but they are handled with such a spirit of refinement and delicacy that they are never offensive. They lean very largely towards exaggerated effects, yet their decorative spirit is tempered with that artistic quality which steers them away from all errors. With them good taste is a natural characteristic. All that they do is the work of artists. They love their country in all its natural beauty, their architecture in its picturesqueness, and their costumes in their brilliancy.

One may go into a Breton kitchen with its dirt floor and a table fastened to legs driven into the ground, and yet see at the end an immense fireplace with graceful outlines, and at the sides beds, armoires, and a clock decorated by the hand of an artist to whose equal we are strangers. We copy his work, but we omit from it that vitality and spontaneous artistic spirit which are always to be found in the architecture of this section.

There is a definite relation between the architecture of Brittany and the country itself. The people, their costumes, the natural scenery, the buildings, and the very atmosphere are harmonious. It has the unity with its surroundings which we so much miss in New World architecture. We must know that these things exist. We must know that there is an architecture harmonious with people of good character, which is a part of a beautiful landscape, in order to realize that we may develop along the same lines and be receptive to the influences which have developed good architecture in other lands.

Here is a street in Auray (Fig. 20). If we should take a section of this city and drop it down in Massachusetts or Ohio, how incongruous it would be! The people, the life, the surrounding country, the very air and sunlight, would render it queer, unnatural, unexpressive of any American thought, and hence inartistic. We may profit by comparing this architecture with that of other countries, and considering our own demands in the light of the general principles which have developed universal successes elsewhere; but our profit will not come through

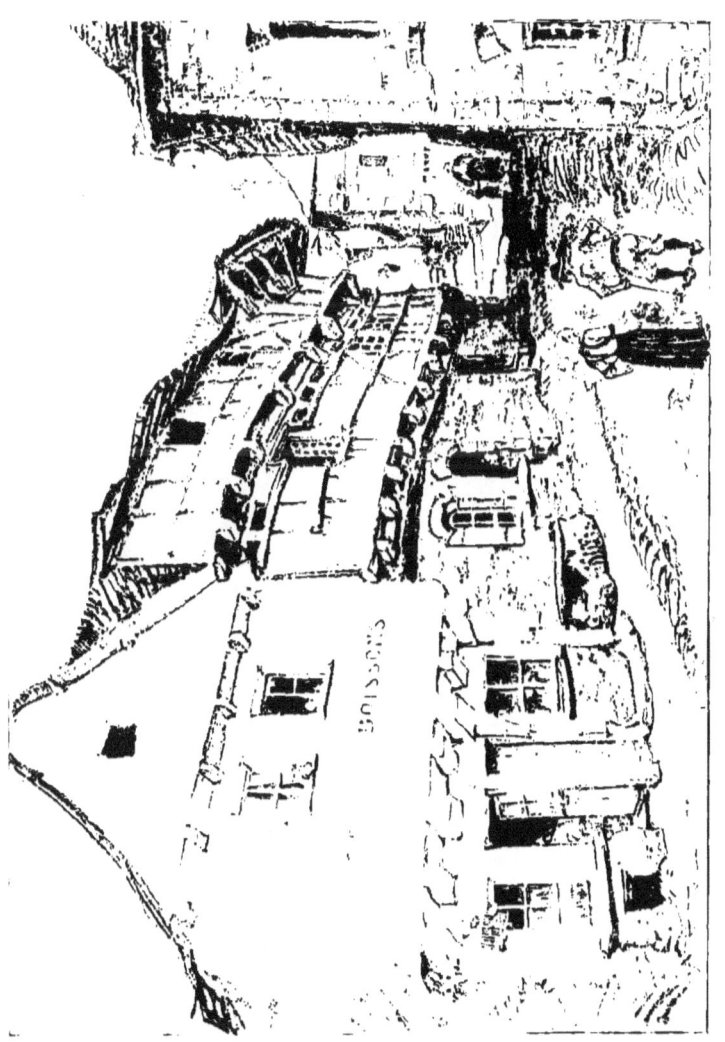

FIG. 40.—A STREET IN ALWAY.
(From "La Vieille France," A. Robida.)

FIG. 21.—IN MALESTROIT.
(From "La Vieille France," A. Robida.)

the literal transportation of Brittany designs to our own States.

I have said that the architecture, the scenery, the people, and the atmosphere are harmonious. This is because of the development of a people in a section naturally beautiful, who have been sensible to that beauty. Whether one meets a Breton at home or abroad, he always speaks of the beauty of his own country. The recognition of this has led him to so build as not to disturb the natural picturesqueness of the country.

The land is undulating, divided into many small farms by means of stone or earth walls which are overgrown by a rich vegetation. Because of the richness and luxuriance of this growth in the parts not cultivated or otherwise cared for, there is something almost tropical in the denseness of vegetation. In the many valleys are small streams lined with tall Lombardy poplars and rapidly growing trees. Although the country is rolling, it is not usually so abrupt but one finds beautiful meadows on each side of the stream; nor are the hills so steep that they are incapable of general cultivation. The elevations are sufficient so that one may often see great stretches of country divided into many small farms by the beautiful walls. The varying colors of the different crops, the stretch of meadows, the long-sailed windmills on every hill, the charming farm villages of thatched roofs, the spires of the beautiful stone chapels lifting their points from among the trees, the calvaries at the roadside, form pictures upon which one never ceases to dwell. The châteaux, standing as monuments to great human events, are often a part of these great pictures.

In walking along a roadway and approaching a town of

any size, one is never disturbed by abrupt changes from country to town or city life. One enters the city of Josselin, with its eight or ten thousand people, its great château, and

FIG. 22.—A SHOP WINDOW IN JOSSELIN.

its church with legends enough to stock a volume, without a perceptible change in sensation. It is this relation of an architecture to the people and to nature itself that makes it so supremely satisfactory to an artist. The details of a building may be never so rich, or the general structure never so pretentious, yet if it cries out against all that is around it, and is distinct in character from those who occupy it, it is not deserving of high consideration.

The view in Malestroit (Fig. 21) is typical of those in

many Brittany towns. The picture shows considerable detail which is common to many buildings. It indicates an utter disregard of perpendicularity in the walls,—a condition not to be imitated in our own work, yet having its proper place here. The street in Auray previously referred to is

FIG. 23.—WINDOWS IN HÔTEL DES VOYAGEURS AT MORLAIX.
(From "La Vieille France," A. Robida.)

another example illustrating the same idea. The little photograph of the shop window at Josselin (Fig. 22) shows a picturesque store-front and some crude efforts in sculpture. They are the master and mistress who founded this store. The street beyond this is quite as interesting as any which have been shown as typical instances of Brittany building.

Fig. 23 is given because it shows more of the detail of these structures than has been given elsewhere. Fig. 24 is

Fig. 24.—Chimney-Piece in House of Anne of Brittany at Morlaix.
(From "La Vieille France," A. Robida.)

a chimney-piece in the house of Anne of Brittany at Morlaix. The tailpiece of this chapter gives a typical illustration of a form of bed very common in this country. This one has a seat at the bottom arranged to form a chest. The bed itself is double, not unlike those used in our modern sleeping-cars. The beds of the master and mistress and the younger children are always near the fireplace in the kitchen, and others are placed around the walls of the room. The beauty of these details never fails to appeal to the artist, not because of the arrangement itself, but because of the masterly manner in which the woodwork of these structures is decorated.

FIG. 25. — A BRITTANY BED.
(Sketch by A. Robida.)

THE WORLD'S HOMES.—*Continued.*

CHAPTER V.*

FRENCH CHÂTEAUX. — MILITARY STRONGHOLDS. — PRINCIPLES GOVERNING THEIR CONSTRUCTION. — HISTORY. — PLAN OF COURCY. — MILITARY BUILDING AND ART. — A BIRD'S-EYE VIEW. — CHÂTEAU OF JOSSELIN. — INTERIOR VIEWS. — PIERREFONDS. — CHAUMONT. — A VIEW OF CHENONCEAU. — AZAY-LE-RIDEAU. — THE SOCIAL CHÂTEAUX.

THE most splendid homes, as well as the earliest in the modern world, were the castles or châteaux, of which those of France are the most notable examples. After Gaul was conquered by the Northern tribes, the victors built themselves châteaux upon the almost inaccessible heights. At the foot of the hill clustered the hamlets of the peasants, who obeyed the master above. This master, in return for their service, gave a sort of protection, one which was at least better than none at all, one which measurably protected them from the neighboring barons, who sought to conquer all within their reach. On the height was a stronghold, surrounded by immense walls. Within the walls were not only the homes of the baron, usually the donjon, but also the quarters for the immediate retainers, the storehouses, which could hold a year's supply of provisions of all sorts, the stables, kennels, a chapel, etc. In time of war the peasantry below were often allowed to seek shelter within

---

* This chapter was written by Emily Gilbert Gibson.

the outer walls of the château, to camp out in the open, which was better than to be left in the hamlet, exposed to the fury of the assailant. But in war, as in peace, the family of the baron usually lived in the donjon, the strongest part of the assemblage of buildings. The donjon was the stronghold, and generally bore no relation to the dungeon. Here, isolated from all but the most trusted servants, the immediate friends, and the servants of the Church, was the first family life of the modern world. Here the wives and daughters worked those splendid tapestries which are to-day a marvel of beauty and artistic workmanship. Here in the evenings the father and sons recited the day's exploits. Thus developed a typically beautiful home life, in which old age was revered, the weak protected, woman loved and cherished. The very institution of feudalism, which later worked so much of woe, gave the best conceivable opportunity for developing the sweetness of domestic life. The strength which could repel the assailant maintained, in the midst of the universal bloodshed of that period, a quiet and repose necessary to the development of domesticity.

Of the earliest châteaux nothing remains now but heaps of ruins. We have documents which give sufficiently lucid accounts to admit of a thorough knowledge of their structure and form. But they were rude structures, with little of the artistic about them, excepting as their perfect adaptation to the need which brought them forth gave them a strength, picturesqueness, and ruggedness, which were both beautiful and artistic.

In the thirteenth century, when the great Gothic cathedrals and churches were built, there were also erected a large number of châteaux, which are both impressive and beautiful.

At the death of Philip Augustus, in 1223, one of his most powerful vassals, Enguerrand III., Sire de Courcy, dreamed of seizing the crown of France. But Queen Blanche forced him to take oath of fidelity to the infant king, Saint Louis, and nothing was left to Enguerrand but to show his power in the erection of a château rivalling the Louvre, wnich had been begun by Philip Augustus. From 1225 to 1230 there was built not only the immense structure, which is one of the marvels of France to-day, but also the walls of the city of Courcy. In Fig. 26 is given a plan of the château and town, in order to show their relation, which was typical. The château B is built at the end of a steep escarpment on the most inaccessible part thereof. The city C is at the other end of this escarpment, and is approached by three roads, which are easily found on the cut, one being at the north, one at the south, and one at the east. A is a fortified esplanade, situated between the city and the château, for in those times the residents of the château had small faith in the fidelity of any one, much less in that of a whole city. D is a place d'armes, separated from the esplanade by a deep moat. The walls which surround the esplanade and the city are perfectly plain.

FIG. 26. — PLAN OF CHÂTEAU AND TOWN OF COURCY.
(From Viollet-le-Duc's "Dict. de l'Architecture.")

Fig. 27 is a plan of the château itself, and is typical of the military structures of the whole of France during the entire feudal period. The only entrance is by the passage

THE WORLD'S HOMES. 63

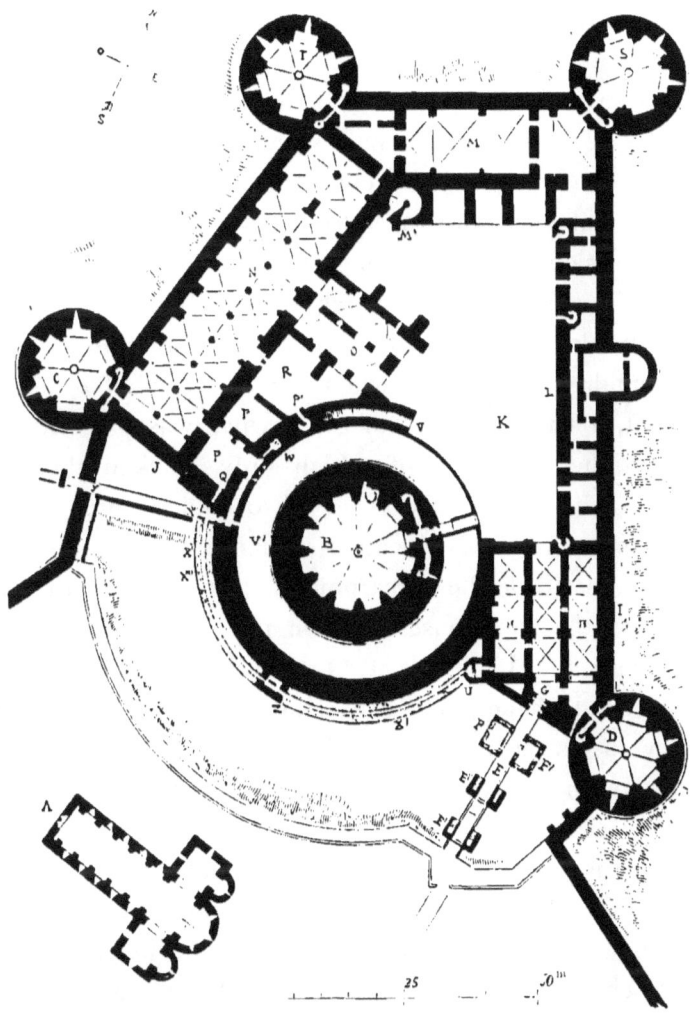

Fig. 27.—Plan of Château de Coucy.
(From Viollet le Duc's "Dict. de l'Architecture.")

E over the moat. In order to gain the point G it was necessary to pass four sets of fortifications, E', E'', F, and the doorway in front of G. The passage from G leading to the court K was guarded on both sides by the two salles-de-garde, II and II'. The donjon B, contrary to the usual custom, was not the dwelling of the family, but was reserved for military operations or for large social gatherings. In it was a large circular hall which could readily hold a thousand men. It was the custom for the baron (standing in a gallery at one side) to address his retainers in this hall.

This donjon was the strongest part of the château, and in case of need the family which lived in the apartments at M could take refuge in it. So strong was this structure that a hundred years ago, when the government desired to destroy it, a large quantity of gunpowder was placed in the basement, the explosion of which only cracked the walls slightly. The family rooms at M occupied three stories. They were given this location because it was the least accessible part of the escarpment. At N, on the first story, were the storehouses, and on the second a large hall. At P were the kitchens, at O the chapel. This château had a second chapel, A, outside the inner walls. This was contrary to the usual custom, but in this case the chapel belongs to an earlier period than the château, which undoubtedly accounts for the innovation.

Fig. 28 is a bird's-eye view of the assemblage of buildings, and is self-explanatory. It is given to show the exterior of the walls. Fig. 29 is a view of part of the inner court.

This château is a good illustration of the statement that any structure built with a strict regard to the purpose which calls it forth is artistic, if it be rationally decorated. On the exterior of Courcy and all other strictly military châteaux there are few windows, and all the decoration is utilitarian. The only decorative features of the exterior walls are the crenelles and merlons at the top, which were placed there

FIG. 28.—BIRD'S-EYE VIEW OF COUCY.
(From Viollet-le-Duc's "Dict. de l'Architecture.")

in order to facilitate the use of missiles and to protect the defenders; yet what could be more picturesque than the walls across the entrance and around the base of the donjon? The donjon itself is decorated with machicolations, through which to drop boiling oil and red-hot stones on the assailant, and with windows and merlons to facilitate the discharge of arrows. At the very top there are four pinnacles which seem to have been placed there only to satisfy the eye. This, again, was an innovation. But even without these the donjon of Courcy would have been a noble and artistic structure.

On the outside all is prepared for war; all is grim and imposing because of its strength. On the façades facing the court, where there was comparative safety, where the family lived and the gentler emotions of life prevailed, there was an architectural expression of these conditions. The view of the court of Courcy shows this, though not so completely as do the châteaux of a later

FIG. 30. — INNER COURT, COURCY.
(From Viollet-le-Duc's "Dict. de l'Architecture.")

FIG. 3. — EXTERIOR FAÇADE, CHÂTEAU OF JOSSELIN.

FIG. 3. — INTÉRIEUR. FAÇADE, CHÂTEAU DE JOSSELIN.

date. The beautiful mouldings along the buildings on the left, the rose windows and other decorative features of the chapel on the right, are all expressive of tranquillity and refinement.

The château of Josselin was once a better exemplification of military operations than it is now. The great exterior wall, the lower part of which belongs to the ninth century, formerly had hourdage at the top through which to drop missiles on the assailant, and the stream which is now separated from the walls by a roadway then washed the foot of the walls. The great dormer windows in the upper part of the exterior have been added within recent years, though they are in the style of the fifteenth century. The wall which shows to the right of the illustration (Fig. 30) formerly continued entirely around the château court. It is now replaced on two sides by a low wall more in keeping with the times.

The interior façade (Fig. 31), built in the fifteenth century, is one of the most beautiful in the world. Weeks of study will not exhaust the variety and beauty of its detail. The whole façade is in the granite of the country, which, as shown, was most elaborately and artistically carved after it was placed in position. The long, low wall is made interesting by the great dormer windows, no two of which are alike, and yet this variety is brought into complete harmony by the skill of the artist.

Two views are given of the interior — one of a bedroom (Fig. 32), and one of the salon (Fig. 33). The walls have been wainscoted and decorated in modern style, the floors are of polished wood, the hangings and furniture are modern. In all respects this old military structure has lent itself wonderfully to the demands of modern living, and the rooms are not essentially different from those we see around us on every side, except in size and the beautiful oak ceilings. The real difference is in the harmony of coloring and the refinement of decoration, and this makes the difference between commonness and beauty. This, however, shows but little in a photograph.

There is no military structure in France around which centres more of historic interest than Pierrefonds, and it is not less interesting defensively and artistically than it is historically. It was built in 1390 by Louis, Duke of Orléans, to hold the territory against Charles of Burgundy, who was then warring against the throne. There were unlimited resources with which to build the structure, and the architect erected one which was impregnable. Many times it was besieged, and succumbed only to treachery. It stood there first a protection to the Crown, but afterwards became a menace, because it was owned by traitors. Then Richelieu, feeling

FIG. 32.—BEDROOM, CHÂTEAU OF JOSSELIN.

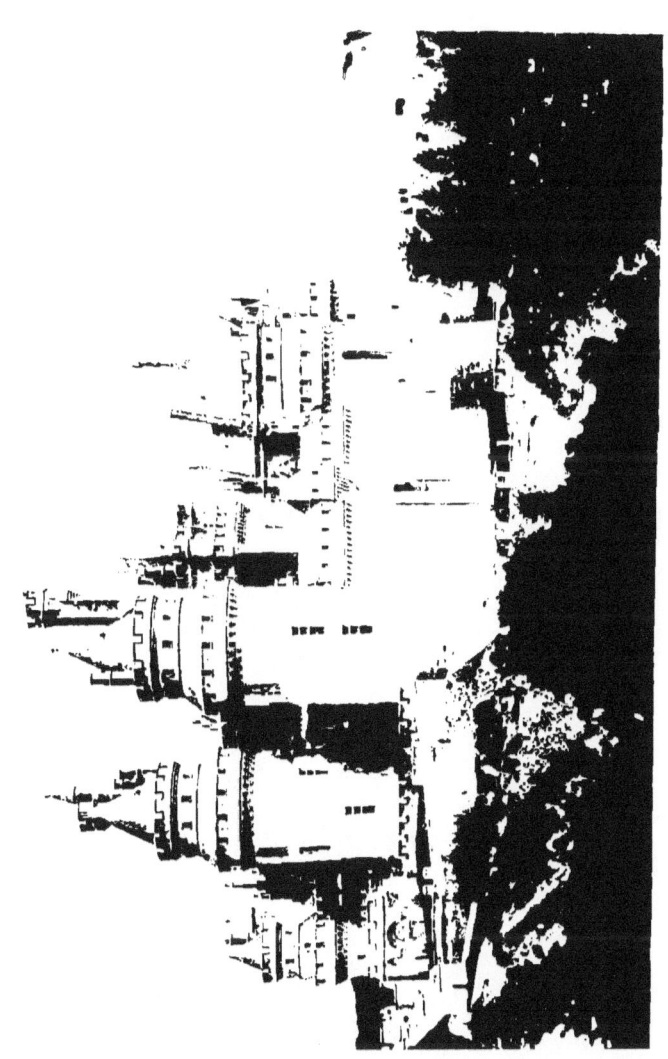

that the government could only be safe with this stronghold dismantled, caused it to be destroyed. For many years the ruins stood there, until Louis Napoleon commissioned Viollet-le-Duc to rebuild it. To-day it stands in all its original strength and beauty — the most complete monument in France. There is not space to give a

FIG. 33.—SALON, CHÂTEAU OF JOSSELIN.

description of either its precautions for defence or the plan of its various parts. In Fig. 34 there is given the entrance to the structure, and in Fig. 36 one side of the inner court. The first shows the majesty and grandeur which followed the adaptation of this exterior to the needs of war. Where there is any decoration it is refined, but nothing but the strictly utilitarian features receive decoration. The view of the interior façade shows the richness of decoration of that part which related to home life. Here imagination ran riot, and there was produced a structure which satisfied the mind and charmed the eye.

FIG. 35.—CHÂTEAU OF PIERREFONDS.

There are many of the old feudal structures which do not adapt themselves to modern conditions as well as that of Josselin. In the old royal residence, Loches, the walls are of the most uncompromising stone, and every expedient had to be adopted to render them homelike from the modern standard. The most successful was in the lavish use of the old tapestries which belonged to this structure. The walls in some of the rooms are fairly covered with these wonderful productions of the needle. The result is grand and imposing. This method of decoration is also

FIG. 37. BEDROOM, CHÂTEAU DE CHAUMONT.

used to a great extent in the Château of Chaumont, two interiors of which are given in Figs. 37 and 38. The exterior of this building follows the general idea of the feudal château, being strong, rugged, and imposing on the exterior, though here there is more decoration on the walls than in the earlier structures, such as Josselin. The interior court is small, and most richly and beautifully decorated. On one side it opens on the river Loire, though it is perched far above it on a crag.

After the sovereignty was well established in France and it was no longer possible for a baron to war with his neighbors, the military châteaux had no further cause for existence. About this time Charles IX. made his expeditions into Italy, and there he and his nobles saw the light, airy buildings of the Italian Renaissance. As a result there was feverish haste in the alteration of the old structures, and in the building of new ones. The walls of the old châteaux were partially torn out, great windows built in, the donjons often destroyed, and everything possible was done to adapt them to the new ideas of living. The result was not always pleasing, for the exteriors of the old structures lent themselves grudgingly to the new ideas of peace and social joys.

The Château of Chenonceau (Fig. 39) is one of the happiest examples of this period. The exterior walls were torn away, the donjon only being left, but entirely separated from the main buildings, the main structure rebuilt on the old foundations, and an entire wing thrown over the river Oise. This latter work was begun by Diana of Poitiers, and finished by Marie de Médicis. To-day the rooms are such as may be found in any large domestic structure where good taste and an adequate sum of money have directed the work. The exterior is unique and most charming.

There is one exception to this general statement. Within recent years the château was restored by a woman of bad taste. As far as the exterior was concerned, the work was left in the hands of a competent architect. The result is good, and the structure stands to-day as left three hundred years ago. But the owner undertook to decorate, according to her own taste, the long picture-gallery which occupies the wing over the river Oise. On this noble room, which is the full length of the wing, untold money was spent, but the result is so bad that it is known as the "Chamber of Horrors." It is almost needless to say that had the same fine sense which directed the restoration of the exterior prevailed here, there would have resulted something of permanent value instead of what must some day be destroyed. The structure is now owned by a native of the West Indies.

Of the new châteaux built in the style of the Renaissance, hundreds of which are to-day found in France, there is perhaps none more admirable than the Château of Azay-le-Rideau (Fig. 40). Here the idea of defence never existed. All is peace, quiet beauty, elegance.

To repeat: In the various châteaux shown, there is not one which is not artistic and beautiful, though with different kinds of beauty. This was brought about through the frank adaptation of the structure to the need which called it forth, and the rational decoration of the motives. In the old military structures it would have been useless and silly to have elaborately decorated the exterior, though as the art

FIG. 58.—BEDROOM, CHÂTEAU OF CHAUMONT.

Fig. 3. — Château de Chenonceau. Revers façade.

FIG. 40.—CHÂTEAU OF AZAY-LE-RIDEAU.

of architecture developed these parts were given refined and adequate decorations. It was always the structural motives which received this carving, and it was always refined and elegant. The history of the people of France can be read in the changes time wrought in the erection of their military structures. In them we see the changes in the methods of war, the development in the art of architecture, the growth of the home idea, the increase in the power of the people, and the corresponding decrease in the power of the barons. But under all circumstances the inherent and studious art of the people is shown. They seem unable to produce a structure lacking in real artistic and refined feeling. On the interior façades of these feudal structures we find enough of art to inspire beautiful building throughout our country, if only we study them rationally and carefully.

A BIT OF DECORATION.
(By Louis H. Sullivan.)

THE WORLD'S HOMES. — *Continued.*

CHAPTER VI.

ENGLISH DOMESTIC ARCHITECTURE. — RELATION OF MODERN TO EARLY WORK. — SELECTIONS FROM HAMERTON'S PORTFOLIO. — DOMESTIC BUILDINGS OF THE SIXTEENTH AND SEVENTEENTH CENTURIES. — PICTURESQUE DETAILS. — FROM THE GOTHIC TO THE RENAISSANCE. — SIX GABLES. — DECORATIVE WOODWORK. — PUGIN'S GABLES. — THE SPIRIT OF DOMESTICITY. — WALTER SCOTT'S HOME. — ONE OF NORMAN SHAW'S INTERIORS.

MUCH of the English domestic architecture of to-day is of the same general character as that which had its origin in the fifteenth and seventeenth centuries in that country. The only material change is in the elaborateness of detail. The general character of composition is practically the same. Modern English domestic architecture is probably more satisfactory than that of any other country. The modern domestic architecture of Germany and France is far from pleasing, and most of our American architecture is very bad. Much of English domestic architecture is interesting, and only a small proportion of it bad; therefore it is fair to say that the domestic architecture of the present time in England is more satisfactory than that of any other country.

In speaking thus well of this architecture it is to be feared that the reader may think that the English architecture is an example for us. While this is true in the broad sense, we must understand that we cannot closely imitate it. English methods are thorough, studious, not over-brilliant, but on the whole pleasing. It is their thoroughness and their quiet methods that we may emulate. One rarely meets the sensational in their domestic structures. Their work is more picturesque than we should expect from the English character. Their buildings abound in gables, irregular roofs, unsymmetrical arrangements of windows, broken wall-surfaces; but all this, which adds to the picturesque, is done so carefully, in so dignified and serious a manner, and withal so naturally, that there is never the least offence. However picturesque their structures may be, they are always dignified.

In the preparation of a short chapter intended to give an idea of the development of English domestic architecture, one labors under the same disadvantage of a wealth of material that presented itself in the consideration of the domestic structures of France. There is so much that is good, so much that might be inserted with profit, that it is with difficulty that a choice is made.

In "The Portfolio," edited by the late Philip Gilbert Hamerton in London, in 1887, Mr. Reginald T. Blomfield published a series of three articles on the half-timber houses in the Weald of Kent and neighborhood. Mr. Blomfield's sketches give, in an ideal way, the spirit of the half-timber architecture of the fifteenth, sixteenth, and seventeenth centuries. I shall make a few selections from his illustrations in order to show the character of work common at that time. It is particularly profitable to observe the difference in detail between this half-timber work and that of other countries to which I have called attention. It is interesting to note how the same style of architecture can be executed in a splendid artistic spirit, and yet vary so greatly in character and detail in different countries during the same period. We should be encouraged at the uniform success and varying character of this architecture. We may learn from this that we can build according to our own needs, and, through serious and intelligent consideration of what we have to do, develop an architecture original in character and of high artistic merit. We can gain encouragement for our own independent efforts from the great variety of work and the uniform successes of this period. The English domestic architecture of the time mentioned followed closely the lines which guided the stone building of that time, though with such changes as would be naturally suggested by the use of another material. At this time oak was very plentiful, and was largely used for the framework and the decorative parts of the domestic structures. The carpenters of this period in England were particularly skilful. There are large numbers of buildings now standing as monuments to their handicraft.

Fig. 41 is a representation by Mr. Blomfield of a Kentish hall of the fifteenth century. The plan in one corner shows the general disposition of the other rooms. This great room was lighted from windows at the sides. It was heated by a fire built in the middle of the room, the smoke passing out through ventilators. These halls were large living-rooms, in which all who were connected with the household were gathered when not at work or asleep. During the latter part of the sixteenth century, and in the early part of the seventeenth, many of these old structures were changed, so that there were two full stories and an attic. This was done by running floors through in line above

FIG. 41.—KENTISH HALL, FIFTEENTH CENTURY

the top of windows. At the same time that this was done brick fireplaces were added to the rooms, so that many

FIG. 42.—A SEVENTEENTH-CENTURY HOUSE AT HEADCORN.

structures of an early date had chimneys dating from the late sixteenth and early seventeenth centuries.

FIG. 43.

Fig. 42 shows a fine old structure of the seventeenth century at Headcorn. Fig. 43 is the spandrel of a tie-beam of the same structure. Doorways of the form indicated by this spandrel, and the class of decoration used in it, are quite common to this architecture. It often happened that one-half of a large doorway was made from a single piece of wood by selecting a tree from which the branches, as well as the main trunk, could be used. This method is suggestive of a naturalness of execution which cannot but lend interest to these structures. Fig. 44 shows a Renaissance residence at Northiam, in Sussex. With a floor plan simple in outline, we yet find the picturesque wall surfaces, which the builders of this time developed in a natural

FIG. 44.

manner. The lower part of this structure is of brick, and the upper part half-timber.

Fig. 45 tells its own story — a number of interesting houses in a row. In the gabled front we can readily imagine that the curved framework was selected from the branch of a tree. Used in this way, it would be a natural form. Builders are now going out of their way to secure such effects. Here the carved branches and part of the trunk lent themselves naturally to such an outline.

In the transition from the Gothic to the Renaissance

period, there was less change in the character of the structures than one might imagine. To be sure, it would

Fig. 45.

hardly be exact to speak of buildings such as we have been considering as Gothic. All that there is of Gothic about them is the detail. There is nothing constructive in this work to give it the stamp of genuine Gothic architecture. However, in many places outside of England there were great changes in outline, even in the domestic structures, during the transition from the Gothic to the Renaissance. In France one often finds Renaissance composition and general outline, while the details are of the Gothic period. This is particularly true of a great deal of the architecture of the early sixteenth century. Much of the decoration — mouldings and other forms — was strongly Gothic, while all of the other features had a decided leaning to the form of composition which afterwards became distinctly Renaissance. The result was somewhat different in English domestic architecture. The transition in style came later in England than in France, and the Renaissance forms were used to decorate the Gothic outlines rather than the reverse, as was common in France.

The long building with six gables to the front, which is illustrated in Fig. 46, is at Bittenden, and is of the date of 1673. Its form and general composition are unquestionably Gothic, though the mouldings and details belong to the Renaissance period. An interesting quality in this

FIG. 46 — HOUSE AT BITTENDEN, SEVENTEENTH CENTURY.

structure is the rigidness of its floor plan and its picturesqueness in external composition. A long, relatively low wall with practically no projections in it would not be the one an American architect would select as the motive for a picturesque structure. The interest is derived from the bold use of gables, and is helped by the variety and character of the framework, brick, and other constructive and decorative features of the walls. It is a straightforward, bold, successful design. The directness with which these gables are used immediately brings to mind the analogous

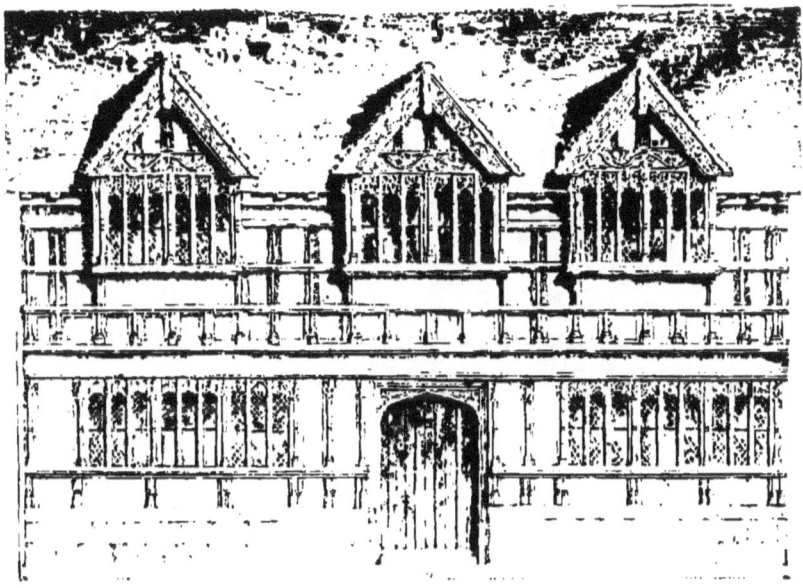

FIG. 47.—FORD'S HOSPITAL. SIXTEENTH CENTURY.
(From Pugin's "Gables.")

composition in the château at Josselin, in western France. There one finds a long façade unbroken in plan, yet rendered wonderfully picturesque in outline by ten splendid dormers.

The sketches just given show only outlines. I shall now reproduce some drawings of detail made by Augustus Pugin, architect, and published in book form in 1831. They are very carefully rendered, and show the rich detail which was common to many of the structures of the sixteenth century.

Ford's Hospital, shown in Fig. 47, was built in 1529. It is a splendid example of the domestic architecture of that period. The mullions and decorative woodwork of the windows and gables are executed in a marvellous manner. The general composition is exceedingly good. Before giving

FIG. 48.—GABLE. FORD'S HOSPITAL.
(From Pugin's "Gables.")

other details it may be well to state that the walls of this structure were of the oak framework shown, filled in with concrete. Somewhat later brick came into more common use for this purpose.

Fig. 48 shows one of the beautiful gables and the carving over the windows of this structure. It is difficult to imagine anything more satisfactory. It was certainly the work of a great artist. Primarily there is beauty of composition in the general structure, and the details are decorated in a most masterly manner. One has a feeling of great discouragement in

FIG. 49.—BOND'S HOSPITAL, EARLY SIXTEENTH CENTURY.
(From Pugin's "Gables.")

the contemplation of a design of this character. There is a delicacy and refined beauty in its detail, a vigor in general conception, which separates this work by a great gulf from anything which is being done to-day. The relation of the windows in the dormers to the lower windows, the doorway, and other parts of the composition is so simple, so elegant, and so masterful that we become despondent after studying it. Fig. 49 is an elevation of Bond's Hospital. It is hardly so interesting in composition as Ford's Hospital, yet it presents detail of the same marvellous character that is found in the other structure. The date of this building is 1506. Figs. 50

FIG. 50.—GABLE, BOND'S HOSPITAL.
(From Pugin's "Gables.")

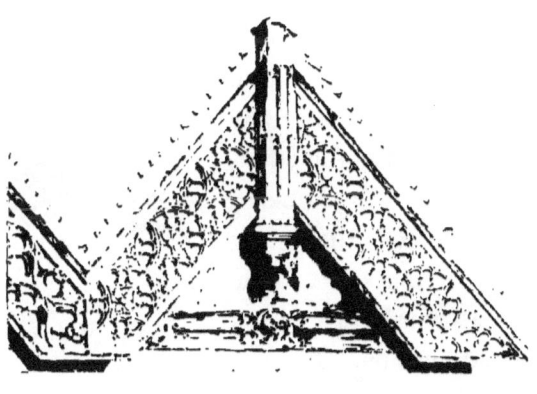

FIG. 51.—GABLE, BOND'S HOSPITAL.
(From Pugin's "Gables.")

and 51 are gables belonging to this building. How, with such splendid examples before it, the world can have gone on building the common houses which are on every hand it is difficult to understand. Yet no doubt in the city of Coventry, which abounds in the remains of this splendid half-timber work, one will hardly find a single structure b u i l t within the last century which will in any way compare with the work of the earlier time. In looking at these designs we cannot but fear that the spirit of the great art-builders of that time has forever left the world.

FIG. 52. — A COVENTRY GABLE.
(From Pugin's "Gables.")

Another of the Coventry gables is shown in Fig. 52. It is hardly so brilliant in design or fine in execution as the other, yet it is a splendid example of the great work of this time. One has no words to express his admiration. All one can do is to look, and ponder over the work of these

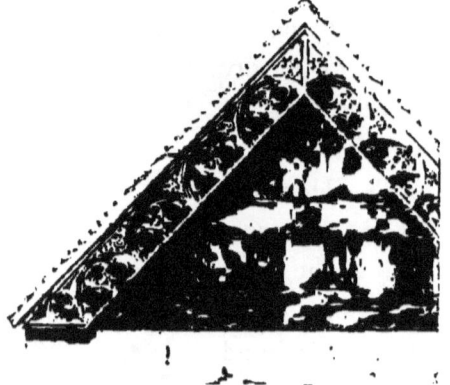

FIG. 53. — A COVENTRY GABLE.
(From Pugin's "Gables.")

great artists. Another of the Coventry gables is shown in Fig. 53.

The spirit of modern English domestic architecture is nearly that of the architecture of our own country. The home spirit permeates the United Kingdom. In this respect it differs greatly from France and Germany at this time. The home as it is known in America is not understood in continental countries. The great establishments of the continent are monumental, impressive in outline, magnificent in conception, but far removed from all ideas of domesticity. The more modest buildings partake of the rigidity and stiffness which belong to their grand neighbors. Either that, or they are apartment houses in which their occupants merely eat and sleep. But homes in the sense known to the English and Americans are not appreciated. The same ideal of homely domesticity that belongs to the cottage runs through the great establishments of the United Kingdom.

The ideal home can be better represented by an example than a definition. One who is interested in domesticity and home life and home building can find nothing more charming and satisfactory than the history of Sir Walter Scott's homes and home life. Everything around him partook of his own sweetness of character, and at the same time was rendered picturesque by those qualities of mind which belonged so much to him. There was the element of romance in everything that he touched. Abbotsford was a narrative, and in it and through it ran the poetry of his nature. But his home-building character had an earlier demonstration than in that completed structure. Abbotsford was its culmination. One of his homes was an old border tower, and in its little sitting-room he once entertained three duchesses. Here he wrote with children, wife, servants, and dogs around him, sitting in the little bow-window which looked out on the Tweed. The first house at Abbotsford was a simple cottage. The ground around it was bare of trees, but was immediately set out with slips and planted with seeds sent by his friends from all quarters of the globe. With the success of his novels Scott changed his plans many times. The building moved along with his resources, but the home idea was never lost during its transition from a cottage to a castle.

One can see in Abbotsford the same idea which brought forth his poems and novels. Antiquarian research formed the basis of both. An old legend, a queer or interesting character, combined with ideas of his own, in his own inimitable way, formed the poem or novel. A block of stone from the door of Tolbooth Prison, bits of carving from Melrose Abbey, carved wood from Dunfernline Kirk, mosaic pavements from Italy, built together with new stones, placed in the form suggested by Scott's imagination and historical research, formed a building picturesque and romantic as the Lady of the Lake. One cannot but feel that Abbotsford was to him more real at times than his literary work. It combined the legends of his country and the romantic, imaginative qualities of his own mind in tangible form.

When we look at an architectural production like Abbotsford Castle, the most natural question to ask is, What style of architecture is it? In truth, Scott's castle is as varied in its style as the legends with which he stored his brain. It is a kind of architectural romance, altogether very picturesque and varied. There are in it suggestions of the

FIG. 54.—A MODERN ENGLISH INTERIOR.
(By Norman Shaw, architect.)

heavy Norman architecture of the twelfth and thirteenth centuries, and alongside this, and at times a part of it, we have Elizabethan details and suggestions of the perpendicular style, and again choice bits of the early English, all arranged in a most picturesque and pleasing way. Thus we have an architectural romance constructed exactly in the same manner as his literary productions.

Fig. 54 is a modern English interior by Norman Shaw.

A DECORATIVE MOTIVE.
(By Louis H. Sullivan.)

THE WORLD'S HOMES. — *Continued.*

CHAPTER VII.

MODERN ARCHITECTURE OF GERMANY. — DRESDEN AND THE ROCOCO. — NORTH OF THE HARTZ MOUNTAINS. — THE OLD ROSE-TREE AT HILDESHEIM. — AN ARCHITECTURE FULL OF FINE SENTIMENT. — IMAGINATIVE ARCHITECTURE. — TILE ROOFS. — CLEAN PAVEMENTS. — DOORWAYS. — NATURAL METHODS.

THE modern architecture of Germany is not especially interesting. One finds there many beautiful structures; but the best modern work is not found in that country. The student of architecture need not feel disappointed if he never see Berlin. In that city are artistic structures, but there are those of the same general character elsewhere which are more interesting. Modern German architecture is largely imitative; there is little that is original. The foundation principles of good architecture are often disregarded. There are in Berlin a number of beautiful modern structures in imitation of the French architecture. There are others which are very interesting in imitation of the old building at Nuremburg. But why go to Berlin to study the early Renaissance of France, or the architecture of Nuremburg? It is infinitely more satisfactory, eminently more advantageous, to study these styles where they originally developed. Berlin, the larger part of Hamburg, parts of Hanover, many of the new cities of Germany, a large part of Vienna in Austria, are unprofitable for study, not

FIG. 55.—BUTCHERS' GUILD HOUSE, HILDESHEIM.

because these cities are not beautiful, but because the same style of architecture which we see there is in a much more satisfactory state elsewhere.

In Dresden we find the rococo architecture in quite as satisfactory a condition as is possible for this style to be. It is arranged exceedingly well; in a word, it is the rococo well done, and is interesting; but we cannot expect to repeat it. We should let it rest as we find it there. It is the flower of a decadence. There is nothing in it for us.

North of the Hartz mountains, in towns like Hildesheim, Halberstadt, Brunswick, and others of that region, we find an early domestic architecture original in character, permanent in construction, and artistic in design. This architecture has a foundation of splendid historical tradition. It was a religious centre as early as the year 822. The churches dating from the eleventh and twelfth centuries are in a good state of preservation. There is now clinging to the choir end of one of the churches in Hildesheim a rose-bush which is over a thousand years old. In the twelfth century when the church was enlarged a bit of wall was built around the bush to save it from destruction, and it has continued to grow ever since. There can be no question but that the sentiment which protected this vine, and which built and cared for the splendid religious structure, developed the artistic enthusiasm found in these people throughout their history.

The domestic and commercial structures built during the sixteenth and seventeenth centuries are decorated with carved inscriptions indicating their fine sentiment.

Oftentimes praise is given that, through the goodness of God, they have been allowed to construct and occupy their houses. On one house is found a lengthy inscrip-

FIG. 56.

tion which reads: "Oh, God, how it always happens that those who hate me, to whom I am doing nothing, grant me nothing and give me nothing, but still must suffer me to live. If they think I am wrong, they would better look out for themselves. But I trust God and do not despair. To them who deserve it, good luck comes every day."

There are many carved panels or decorated sills which deal with mystic subjects, tell stories with a moral, picture the saints, extol the virtues, and in other ways bring to mind the dominant qualities of this God-fearing people. Again we find coats of arms, mottoes, inscriptions, carved images of the occupants and builders of these houses. Altogether, there is much that is religious, mystical, romantic, narrative, and personal in the decoration of their buildings. All has been influenced by an impulse so poetic that the decorative work is devoid of realism. It is teeming with imagination. The stories which are told

FIG. 57.
(From Laechner's "Holzarchitectur.")

FIG. 58.
(From Laechner's "Holzarchitectur.")

through the heart and hand of the wood-carver indicate clearly enough the character of the people. Yet it has none of the directness or rigidity of a historical record. This work is decorative in the highest sense. It is truthful, romantic, and beautiful. In looking over the selections from the architecture of this region, one may note with interest and profit that the decorative parts which have this narrative and emotional quality are not separated from those which are altogether conventional in their form. One may find a narrative in wood in immediate contact with decorative forms which are strictly geometric and architectural. Yet the decorative spirit so closely unites these that there is nothing disturbing in the composition.

FIG. 59.
(From Lacchner's "Holzarchitectur.")

The frontispiece (Fig. 55) of this chapter is of a butchers' guild building in Hildesheim. Fig. 56 is a section of the detail of that structure. This building is considered the finest timbered structure in Germany.

FIG. 65.—HOUSE IN HILDESHEIM.

Fig. 57 is illustrative of an opening, circular in outline, secured by means of braces on each side. Beautiful decorative forms are shown on the upright posts; on the right over the opening are the busts of a master and mistress of the establishment suggested in low relief. Fig. 58 indicates the corbel projection where the decoration is suggestive of Gothic detail of a late type. Fig. 59 is given for the purpose of showing the variety of form and detail which could be used to decorate the same constructive features; viz., the corbelling of the successive stories one beyond the other. Thus while we find in this part of Germany buildings of the same general character, all having constructive features in common, there is never an impression of monotony. The photograph (Fig. 60) shows buildings constructed during the latter part of the sixteenth century in a manner sufficiently clear to indicate the treatment of detail as well as the roof forms and other features of interest. In this work there is more suggestion of the detail of the Renaissance than in most of the other illustrations which have been given in this. All of the roofs of the buildings of this period are of tile of the pattern shown in this picture.

I especially wish to call attention to the clean pavements shown in this photograph, as well as in that of the butchers' guild. This is the universal condition of the pavements in this part of Germany. A town is never so poor or so small but that the streets are well cared for.

The satisfaction in selecting architectural illustrations from the cities named is that any one of them represents the general character of the buildings of that region. One can hardly go amiss in making a representative choice.

FIG. 61.
(From Laechner's "Holzarchitectur.")

There is very little difference in the degree of artistic excellence. The older parts have size, beauty of outline, richness in sculptural decoration, and wonderful coloring to make them beautiful. The structural forms are simple enough. A curved line which acts as a supporting brace is found only on the first floor. The sills for each story project beyond the walls, and these, in turn, are supported by decorative corbels or brackets. The upright posts continue from the lower sill to the one next above it. They, in turn, are braced to the sills.

An examination of these buildings will indicate clearly enough that we could not displace any of the decorative features without removing a vital part of the construction. All is decorative construction; there is no constructed decoration. In this is realized the highest state of architectural art. Figs. 61 and 62 indicate as clearly as possible some of the characteristic forms and constructive features of this architecture. Fig. 63 is an interesting doorway in wood, and shows the decorations of the sills, other forms of radial

FIG. 62.
(From Laechner's "Holzarchitectur.")

decoration, and the manner of handling an inscription. A hasty judgment of this doorway would be that an arch form in wood is not justified. But if this construction be examined more carefully it will be seen that while the opening is circular the construction is not an arch, but rather a brace. No effort is made to suggest the arch, while the brace lines are clearly indicated. The decoration of the chamfers, or the projecting sills, and that of the brackets suggests a knowledge of Romanesque forms. These could clearly come from the old churches of the eleventh and twelfth centuries, in which this region abounds. These cities con-

FIG. 63.—A DOORWAY OF CARVED WOOD.
(From Lauchner's "Holzarchitectur.")

sist entirely of buildings of this character. There are differences to the elaborateness of the detail, but never an unsuccessful structure. All are beautiful. While they are essentially different from the buildings of other regions referred to in this book, there is no deviation from the mark of high artistic excellence. This section is original in its architecture at the same time that it sustains a high artistic standard. It is for this purpose that the selections from France, Germany, Holland, Switzerland, England, and the early architecture of our own country are made. Each is entirely distinct in character, yet there is no great variation in merit. We may find encouragement in the fact that we may be quite as original in the work which we have to do. Neither the field of originality, nor the possibilities of artistic excellence are exhausted. We may follow the natural methods of our own time, and yet under the proper artistic impulse develop a domestic architecture quite as interesting as any the world has known.

Fig. 64 is characteristic of the architecture of South Germany.

FIG. 64.—A MODERN SOUTH GERMAN HOUSE IN THE STYLE OF THE SIXTEENTH CENTURY.

THE WORLD'S HOMES. — *Continued.*

CHAPTER VIII.

SWISS ARCHITECTURE. — RELATION OF SCENERY TO A NATION'S BUILDING. — NATURAL DOMESTIC EXPRESSIONS. — ARTISTIC FORMS. — NATURAL COLORING. — VARYING METHODS. — CHARACTERISTIC ILLUSTRATIONS.

SWISS domestic architecture is another natural independent artistic expression. It has a decided style of its own. Its character is distinctly marked. Its construction is honest; the decoration is the work of an artist. It houses an industrious, serious, provident, home-loving, independent people. There is something about a high sentiment which makes great art. While Holland was struggling for existence she was producing great paintings and artistic buildings. The emotional, religious sentiment of the Middle Ages produced a universally brilliant architecture. The stolid seriousness of the German, the quiet progressiveness of the English of the sixteenth and seventeenth centuries, and the pronounced domesticity of the Swiss, each found expression in their architecture. Our American architecture must come in the same way, as a sincere, natural expression; it cannot be forced. It cannot carry with it our mannerisms, our affectations; it will not come through mere effort. Hence, it may be slow of development. The fact that it becomes less easy to earn money as the country grows older will help the progress of art. It makes people more thoughtful in their expenditures, and added thoughtfulness in any direction is helpful. The Swiss peasants, workers, and merchants did not have in mind that they were producing great art. They were not seeking to be individual. They did not build broad, flat-pitched roofs because they harmonized well with their scenery. This was the kind of roof best suited to their surroundings. However, one cannot doubt that the wonderful scenery created the love for the beautiful which unconsciously developed a style of decoration and forms of construction in keeping with the character of the people. These people have now, as they always have had, a sincere affection for their country. It is a high sentiment such as always develops good art.

While there is a general character belonging to all Swiss houses, it is true that each canton has its peculiarities of decoration and methods of construction, and hence, in a way, a distinct style of its own. In one section we find a half-timber construction, which, considered constructively alone, is not greatly different from that of some of the South German houses. Yet its decorative character gives it a distinct type. These half-timber houses are filled in with stone masonry, and are plastered and whitewashed. The woodwork is dark-reddish in color and makes a splendid harmony with the white walls. Another class of Swiss building is allied in principle to blockhouse construction. We see some of it in our log houses of recent years, though with us there has never been an attempt at decoration. There, however, is similarity in constructive principle. In these houses the timbers are dressed on all four sides, and the corners notched together with projecting parts of timber. These projections form brackets, consols, pilasters, braces, all of which are beautifully decorated. An example of this architecture is shown in Fig. 65. The natural forms, the rich, beautiful decoration, may well call for great admiration. The house is deserving of careful, thoughtful study.

FIG. 65.—HOUSE AT MEININGEN.

We may emulate the spirit if not the methods and details of this great work. The deep shadows cast by the projections serve to preserve the natural color of the wood in the parts best protected. Where the shadows are the strongest we have a rich, natural red, and as the sun and storm reach the other parts we find various shades of gray and brown, all forming a perfect natural harmony. The covered passages of these structures form picturesque and interesting features.

Another type of Swiss architecture is a kind of heavy

FIG. 66.—HOUSE AT MEISINGEN.

plank construction. It has a typical illustration in Fig. 66. This example indicates as well as possible that the decoration follows the true nature of wood. The illustrations given are of houses of the seventeenth and eighteenth centuries. Their lasting qualities and their artistic beauty show the fallacy of our use of paint as a wood preserver.

There is something in the spirit, though not in the detail, of this construction and decoration which is suggestive of the Japanese.

THE WORLD'S HOMES. — *Concluded.*

CHAPTER IX.

OLD COLONIAL ARCHITECTURE. — A CLASSIC DEVELOPMENT. — CHARACTERISTIC NEW ENGLAND ARCHITECTURE. — LUXURIOUS CHARACTER OF THE OLD COLONIAL IN THE SOUTH. — THE BEST AMERICAN EXPRESSION OF DOMESTICITY. — ILLUSTRATIONS.

THE term "Old Colonial," as applied to architecture, is broader in its significance than when applied to a geographical section — the old Plymouth Colony. The Old Colonial architecture was a distinct type which had its development in the newly settled sections of our country during the eighteenth century. The early examples of this work are to be found along the coast line from Massachusetts down the South Atlantic States, across the Gulf, and up the Mississippi river. Thence this style of building reached the interior, so that now we find many examples in the States immediately east of the Mississippi river. The severest type of this architecture is found in New England; in Maryland and Virginia it is quite sumptuous, and through many of the South Atlantic States this same character is preserved. The earlier architecture of the lower Mississippi, New Orleans, Natchez, and Baton Rouge shows the French influence, being more monumental in character and richer in detail, and in other ways indicating its Latin origin.

In Old Colonial architecture are used classic forms in the decoration of domestic structures. The architectural work of New England was largely done by educated English carpenters. It was careful, thorough, academic housebuilding. This is the best known example of an architectural expression of the character of a people. The old New England house and New England history are variations of the same expression. It may be that future generations will say that the heterogeneous domestic architecture of our own time is the truthful expression of a heterogeneous people. But this is not for us to say. We look upon the early New England character as thoughtful, refined in a way, and very severe and exact. We find the architecture of that period expressing as clearly as possible all of these characteristics. There is no suggestion of the picturesque either in the personality or the architecture. In this architecture there is no mere prettiness. It is truthful and refined. It carries with it a certain dignified quiet beauty and nobility which always come from concentration. How rarely can we say this of our modern domestic work!

In Maryland, Virginia, and the South Atlantic States this early work assumed richness of detail and often a suggestion of the picturesque — an architectural quality altogether in keeping with the livelier character of the people. Through the slave States, notably in Kentucky, we find splendid mansions occupying large plats of ground in the cities and dignifying the highways as the centres of great plantations. This architecture adapted itself to the dignified living of the wealthy slave-owner. In Mobile the early merchants built their mansions on Government street and other splendid driveways, and we have there many magnificent establishments. A great deal of the work along the lower Mississippi was planned by the French architects and executed by imported artists. The plans of these structures conform admirably to the domestic and social life of the times. Go where we may, we find this Old Colonial architecture impressive and full of stately beauty. The dignified social intercourse which we are taught to believe belonged to that people and period appears to have been a part of the general organization which developed their architecture. In the country mansion of the Southern planter we find the great porch in front with its classic columns extending through the two stories of the house, forming what is known as a gallery on the second floor. Running entirely through the house is a wide hall, with a long double parlor on one side, the wall of which is decorated with classic pilasters and cornices. The richly decorated columns of one of the classic orders project from the wall on either side far enough to divide this long room. The woodwork and plaster are usually painted white. Rich fireplaces, large mirrors, family portraits, busts, and bits of statuary decorate the wall. In the French section of the South, the furniture of these establishments usually came from France. We find in New Orleans to-day second-hand stores well stocked with the furniture and fittings of these early magnificent establishments. The external details of these buildings were classic in form, and designed by well-trained artists.

The fittings of the New England structures were much quieter in character, always in good taste, yet rather severe and prim. The material was good, the form graceful, but there was little that was decorative in the way known to the Southern architecture. The typical New England fireplace is well known to all of us, not only through reproductions in the buildings of our own time, but by pictures from the earlier models. It is pleasant to associate the clean-cut mental products of early New England with these architectural emblems of quiet thoroughness. This early work fulfils all of the conditions of good architecture. It is expressive in a beautiful way of the character of the people of its time.

In Fig. 67 is a typical example of an Old Colonial mansion. In this instance, the great Doric porches, two stories in height, are at either side of the house. Fig. 68 is the floor plan. This example is taken from Edward Shaw's "Rural Architecture," published in Boston by James B. Dow in 1843. This architect gave plans of interesting houses of this general character. In some instances he took a floor plan and gave designs for the exterior in the various orders — the Doric, the Ionic, or the Corinthian. But most of all the book abounds in carefully constructed details, large scale drawings of the various parts, full-sized details of the mouldings, and explicit directions for producing the nicety of detail necessary to the artistic development of his work.

FIG. 67. — ELEVATION.

FIG. 68. — FLOOR PLAN.

It is only through such care and thoroughness that we may expect to develop a successful art.

Minard Lafever published a book entitled "The Beauties of Modern Architecture," in 1839. Of this work he said: "Notwithstanding the many works which have hitherto been published on the subject of architecture, there has none yet appeared intended exclusively for the operative workman. It is therefore thought proper to present to the industrious and ingenious a book of original designs," etc. He then proceeds to give a large number of details of doorways, windows, cornices, and other decorative features, which he has developed in the spirit of a true artist. In making a few selections from his work, it is much more difficult to determine what to take than what to omit.

Fig. 69 is of a front doorway, as chaste an example as one can well imagine. This illustration was followed by a large drawing showing the exact form of even the most insignificant mouldings.

Fig. 70 shows one method adopted by this artist of treating the design of double parlors. There is a great temptation to make other selections to illustrate the great care exercised at this time in developing the fine details necessary to artistic success in any direction. Both of the architects referred to use principally Greek details in their work.

FIG. 60. — DOORWAY.

Both Greek and Roman were used in the Old Colonial work.

There are many workers in this style at the present time, but it is rare indeed that we find work that does not show the influence of the crude architecture which is on every hand. It seems that educated architects should be strong enough to avoid such influence, but it is rare indeed to find select work of this general class executed to-day which does not show in its composition and many of its details the influence of the unstudied and carelessly executed work which surrounds us. The writer has known many young men educated in architectural schools where only classic architecture is taught, and has noticed that nearly all of them succumb to the influence of common environment. Modern work of this character developed with anything like the fidelity necessary to good art is exceptional. The example of modern work given (Fig. 71) is much better than that seen on every hand.

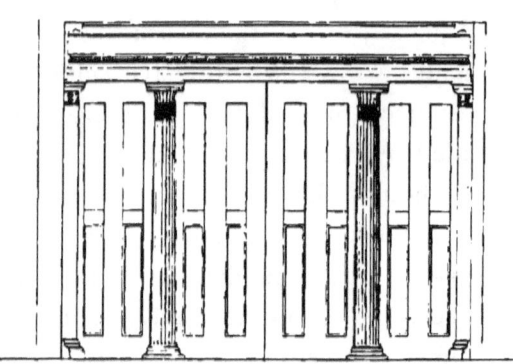

FIG. 70.— DOORWAY BETWEEN PARLORS.

FIG. 71.—A MODERN EXAMPLE.
(Willis K. Polk, architect.)

SOME HOUSE PLANS.

CHAPTER X.

A BRICK HOUSE WITH A STONE FOUNDATION. — FOR A NARROW LOT. — THE PROBLEM OF LIGHTING. — A NEW CLOSET. — RELATION OF THE EXTERIOR TO THE LOCATION. — THE DORMERS. — THE INSIDE FINISH. — MANTELS.

A BRICK house with a stone foundation and a slate roof is an expression of solidity that appeals to many people. Fig. 72 is such a house. The outline is carefully considered, and while the details are relatively simple, they are designed with the greatest care. in order to make the most of the limited opportunities afforded.

The floor plan is somewhat conventional, and not unlike many which are found throughout the West (Figs. 73 and 74). How-

FIG. 72.

ever, the effort was made to keep it simpler than is often the case even in a plan with an origin as straightforward as this one.

The lot is a narrow one. There is a tall dignified house on the south side, and a cumbersome brick structure on the north which approaches the north line very closely. Both of the adjoining houses were set within fifteen feet of the sidewalk. It was necessary to place this building on a line with them in order to give a view up and down the street, and also to prevent its having that receding appearance which always follows the placing of a house farther back on a narrow lot than its immediate neighbors. It was necessary to consider the lighting of the middle rooms very carefully. The other buildings being so close to it, every advantage in the way of light had to be accepted. As is shown by the position of the stairway (Fig. 75), the hall window is particularly large. The dining-room window is broad, though it is five feet from the floor. The windows on the side are of the ordi-

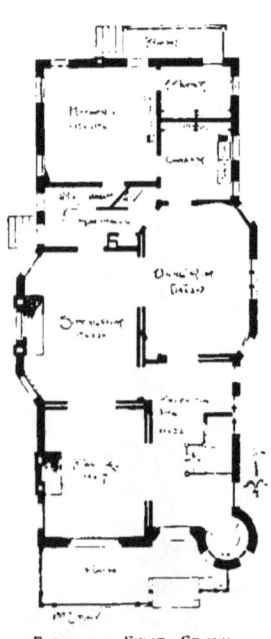

FIG. 73. — FIRST STORY.

FIG. 74. — SECOND STORY.

nary kind. In the sitting-room there is a large window over the mantel, which is indicated in the photograph of the side of that room (Fig. 76).

The relation of the kitchen to the rear porch, the pantry, cellar, china-room, and dining-room, is clearly shown. In the side hall there are doors opening into the sitting-room, dining-room, and china-closet. There

FIG. 75.— LOOKING FROM THE PARLOR.

is a wash-stand and cloak-closet in this room. The laundry is in the basement. The heating-apparatus is located under the sitting-room.

The problem of lighting the second floor is not so serious as the first. The windows are nearer the roofs of the adjoining houses, and for that reason nearer the source of light. The bath-room is separated from the water-closet room by a door. Each has an independent connection with the hall. In the attic stairway there is a clothes-chute to a closet in the basement. The closet on the attic stairway is intended for brooms. There is a linen-closet in the front hall. The closet shown in the alcove needs some explanation (Fig. 77). It has always been a thought of the writer that a closet could be arranged in the same syste-

FIG. 76. — THE SITTING-ROOM MANTEL.

matic way that is common to desks — a place for everything. A great many things are hung up in closets, placed in a miscellaneous way on shelves, or laid on the floor, for the want of a better place or a better way of disposing of them. Dress-skirts and waists and other apparel can be better cared for spread out in trays or shallow drawers than hanging from one or more points of support. This is also true of men's clothing. The advantage of compartments, or pigeon-holes, if they may be so called, for shoes, bonnets, and hats cannot be questioned. Above the drawer-space there is sufficient hanging-room for all purposes, though with the other facilities little hanging-space is needed. The right kind of attention has not hitherto been bestowed upon closets.

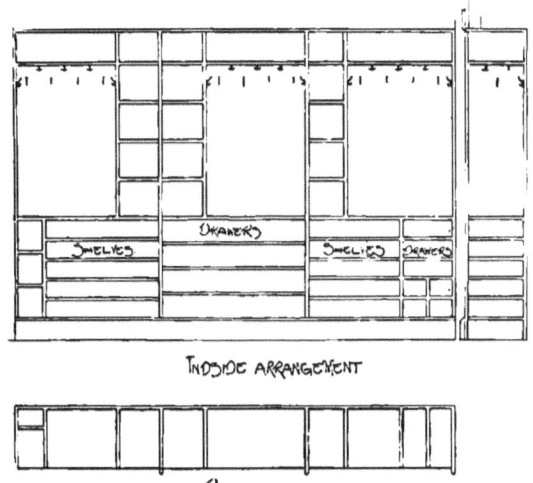

FIG. 77. — DRESSING-ROOM CLOSET.

In the attic there is a large servant's-room back of the stairs, and the great finished attic-room occupying all the space to the front.

The general thought in the design of this structure, after the practical considerations of arrangement and light were considered, was to lift it out of, and away from, its surroundings. It was necessary to do this without giving it great height. The dignified

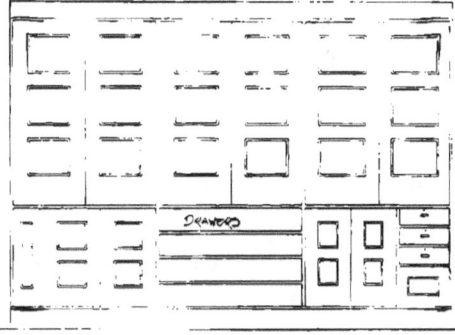

FIG. 78. — CLOSET WITH DOORS CLOSED.

character of the people for whom it was designed did not admit of accomplishing this by a resort to brilliant means. The design had to be strong, graceful, not brilliant in color, yet with a sufficient buoyancy and refinement of detail to lift it above a depressing surrounding. This suggested a quiet, not greatly projecting tower on the corner—one with no strong horizontal line until it reached the top; a vigorous, upright stem. The corbelling is of stone, the arches and material above of brick. The detail is subdued, in order that it may be viewed in the mass rather for any particular feature. For that reason even the finial was kept very quiet. The strong upward movement is preserved in the dormer. In the gable next to it there is no marked division between the brickwork of the gable itself and the wall below.

The caps of the windows are of brick, so that the color of this part need not be disturbed. The woodwork is painted a dark reddish-brown, in order that the eye may not be suddenly arrested in its upward movement from the bottom to the top of this structure. Had the porch cornice been carried straight through, it would have neutralized this effect. However, with its carefully made tracery designed rather brilliantly, this part of the structure indicates to the world that there is nothing disheartening even if one is somewhat crowded. It gives the building a very cheery look under the circumstances. A heavy porch roof coming down over a porch without an opening in the gable is often depressing to one who occupies the porch, though the occupant may not know exactly what is the matter. The tracery of this porch admits light into the front window at the same time that it presents a very pleasing silhouette from the parlor.

The side gables are designed in about the same spirit as those in front. It is the purpose of these, the dormers, and the roof to enliven the whole structure, and make one forget the heaviness that is all around. The cheeriness of this is heightened by the easing of a curve at the foot of the rafter which rounds out the slant of the roof before it reaches the cornice moulding. This is a universal feature in the roofs of Brittany, which lends much to their interest. They are always graceful. It is pleasing to see the slant of a roof end with this little sweep at the bottom. It does away with all of the rigidity of a direct and formal termination.

The side dormers are entirely of slate. The idea in using this material in this part of the roof was to avoid the unrest which would come from the use of a great variety of material. The use of slate secured the picturesqueness of outline without sacrificing the repose, which must have been the result had a great variety of material of varying color and texture been used.

So much for the exterior. After the plan was made and the lighting provided there was no great difficulty in treating the interior. It was merely a question of the refinement of detail. For instance, the pattern of the stairway is rather pronounced, and it had to be developed in a fine, delicate way to prevent its becoming gross and aggressive. The members are small, the mouldings soft, the recesses light. A pronounced outline may be kept in subjection in this way. It is simply a question of refinement in little things.

Leaded-glass is in the large window above this stairway. The glass is all of one color, a soft amber. While a large volume of light was desired, direct white light was not wanted.

FIG. 79. — STAIR WINDOW.

The view of the outside of this window (Fig. 79) shows the careful designing of the stone details. The label moulding over the arch, the finial, and the lines surrounding the window itself as they come to the sill, are made with brick moulded with the corners cut off, as is shown by the section. This is much less harsh in effect than would be a direct recess on the side.

The top of the dining-room window, which is above the buffet, is of the same general form as the one just mentioned. Its sill is on a line with the top of the wainscoting, which is five feet high. The glazing of this window is of the same character as that mentioned in the hall; that is, cathedral glass of quiet color, set in geometrical leads.

The sitting-room mantel (Fig. 76) has a very broad breast — eight feet. Nearly all of the face below the shelf is of tile, only a narrow margin of wood being used on the corners. The woodwork above the shelf extends to the ceiling.

The mantels of the second floor are exceedingly simple. They are faced with white enamel tile. The doors and casings of the first and second floor are given as much for the purpose of showing their simplicity as for any other

SOME HOUSE PLANS. 143

reason. All of the woodwork used in this structure is of quartered white oak, excepting one north bedroom, which is of bird's-eye maple.

Figures 80 and 81 are doors in this house.

FIG. 80.

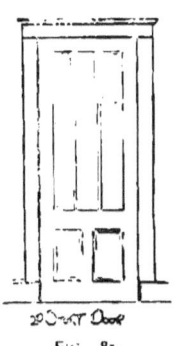

FIG. 81.

SOME HOUSE PLANS. — *Continued.*

CHAPTER XI.

A CENTRE-HALL PLAN. — FRAME BUILDING. — A LITTLE ROOM FOR CLOAKS AND WRAPS. — DECORATIVE FORMS. — INTERIOR PHOTOGRAPHS. — EXTERNAL DETAILS. — GREEK MOULDINGS.

NEARLY every one tries to build for less than the cost of the material and labor which go into the building. While this thought is not definitely formulated, there is a fixed mental limit in cost and a maximum demand in requirement which represent a value in excess of the cost limit. Naturally there has to be some kind of a compromise before the building is erected. It is true that one person may get a much better house for a given sum than another, yet this is not because he gets the house for less than it costs, but because he has a good design. A poorly designed house is worth less than it costs. The material which goes into it is put into

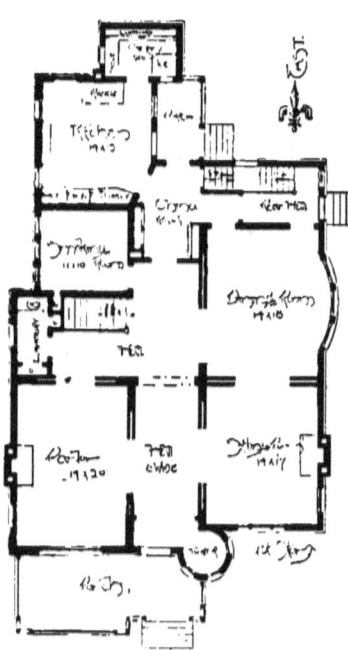

FIG. 82 — FIRST STORY.

inconvenient or ugly forms. Manifestly this same material and labor might be put into a convenient and attractive form and cost no more money. Thus, one who builds from a good design carefully conceived and executed, gets more for a given expenditure of money than another, because he gets the same amount of material and labor into better form. A good plan is always in style. The rational architecture of the thirteenth century is in better form to-day than any which has followed it, because it was designed in a high spirit of rationality, and executed by artists. Good taste is perennial.

The plan given in Figs. 82 and 83 is a house with a central hall, a general type of plan which commends itself to the majority of people interested in housekeeping. A central hall carries a feeling of sumptuousness, freedom, and proper isolation which is not associated with other plans.

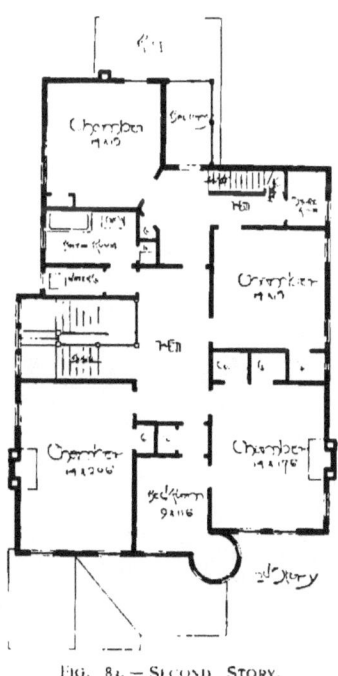

FIG. 83.—SECOND STORY.

The reception-hall idea has very often been carried beyond the reason which gave it birth. It was brought into prominence and popularity as a relief from the old side-hall plan. These halls were mere passages. Usually they were not more than seven feet wide, and rarely less than twenty feet long. Seven times twenty are one hundred and forty. Twelve times twelve are one hundred and forty-four. The architects said: We will give about the same area that was in the old side-hall, and you will have a better room. This is a splendid thought if rationally developed. But where the reception hall is made very large, with a front room

FIG 84.

separated from it by a mere screen of spindle work, the whole front of the house is practically stair hall. As ordinarily arranged, one can see a hat-rack and all that goes with it from any part of these two rooms. This reception-hall idea, when carried to an extreme, has

SOME HOUSE PLANS. 147

FIG. 85.—LOOKING FROM PARLOR TO SITTING-ROOM.

unprotected openings into a room which, merely for the purpose of designation, is named a library, or to other rooms, the openings to which are often filled with ropes, strings, or spindle work, so that in fact there is only one great room on the main floor, other than the kitchen and dining-room. This is the extremest form of the reception hall, and it makes a second-floor sitting-room a necessity. It is an artificial development made possible by the class of people who are always abreast of the fads, and who operate from impulse rather than reason.

The house with a centre hall has a parlor on one side of it, a sitting-room and dining-room on the other, and back of the stairway a little smoking-room, library, or work-room. The rooms can be separated from one another or the hall by

FIG. 86. — LOOKING INTO STAIR HALL.

sliding doors. Then the sitting-room may be a sitting-room in fact, the parlor a reception-room in the rational acceptance of that term. One can live on the first floor of such a house and not disturb the various movements of household affairs. In the front part of the hall there is a little vestibule, and connecting with it a circular tower which forms a room for cloaks and wraps. The hall is divided by two lines of carefully designed fretwork. The patterns are taken from old Byzantine floors of the sixth century. They are carefully developed in oak, and give very interesting silhouettes against the light. The sliding-door openings are decorated in the same way. Details of these are shown in the sketches which accompany this plan. While the same general style of pattern is used in the doorways as in the hall decoration, there is an anachronism in the use of the fleur-de-lis and the ermine as the central figure of the geometrical Byzantine outlines. The mantel in

FIG. 87. — FROM PARLOR TO STAIR HALL.

the sitting-room has the outline of some of the old Dutch mantels, and the decorations are of the general character of those mentioned elsewhere. A purist might object to

Fig. 88. — Sideboard.

these mixtures, and say, Why not take one style and follow it? Yet why follow it? Some of the best architecture that the world knows is a mixture. The best things have not been done by following decorative motives merely because they belong to a particular style. In looking at the designs for the decorative woodwork which has been described and illustrated, one would hardly discover that its origin was Byzantine. Yet by comparing the designs with the originals one can readily determine their source. The adaptation is free, as it should be. In this instance it is successful. The stairway decoration is of the same general character.

The relation of the various rooms in the back part of the house is clearly indicated on the plan. The stairway to the cellar is a very convenient arrangement. It goes down to a level six inches above the grade, at which point is a door to the outside and a landing the full width of

FIG. 86. — PARLOR MANTEL.

the door. Thence one can go to the cellar. Thus it serves the double purpose of an inside and outside cellar-way, and has a decided advantage over the ordinary outside flat cellar-door.

The second floor is planned according to the general principle which controls all of the plans of this book. The servants' room is separated by a door from the front part of the house; the rear stairway continues to the attic. The water-closet room and bath-room are separate; there

FIG. 90. — A BEDROOM MANTEL.

is a clothes-chute from the bath-room on the second floor to the cellar; in the rear hall is a large independent linen-closet; there are two closets for the principal front room, and one large closet for each of the others; the attic is finished in one large room over the body of the house, and a tank-room over the rear part.

The laundry is in the cellar, which also contains the servants' water-closet and the water motor which lifts cistern water to the attic. The heating apparatus is under the centre hall, and is provided with an electric regulator which maintains a uniform temperature in the house throughout the winter.

The exterior of this building is a perfectly simple expression of the general internal arrangements. The floor plan was made first, and the exterior is its natural clothing. Its distinctive character is secured solely by the character of the detail which decorates the various constructive features. With respect to the external-wall construction, it may be said that the foundation walls reach to the under side of the first-story window-sills, and thus show an added height of foundation. The upper walls are formed by covering the studding first with seven-eighths-inch flooring, then with paper, and finally with the outside covering for the first and second floors, in the manner indicated on the sketches (Figs. 91 and 92). The sizes of the various materials are shown.

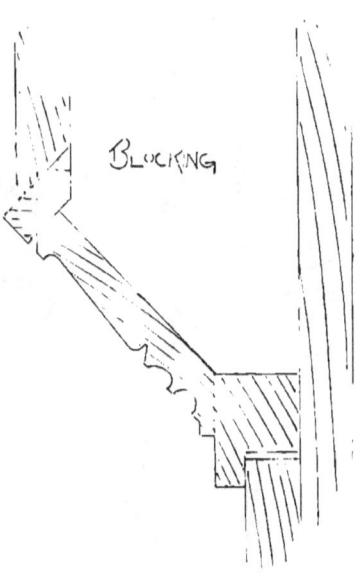

FIG. 91.

The mouldings of the various parts are constructed on

FIG. 92.

the same principle as those referred to in connection with another structure. That is, the Greek idea of light, soft curves with sharp, fine incisions to catch the light, is followed as closely as possible. While the general form of a dormer or the tower might be that of a château of the sixteenth century, they are decorated with Greek mouldings.

FIG. 93. — SITTING-ROOM WINDOW.

A separate larger photograph of the front sitting-room window is given to show the treatment of the wood decoration at this point.

On the north side of the house there is a large window the full width of the stair hall, with about the same outline as the window in the sitting-room (Fig. 93). It is glazed with two light tints of cathedral glass, and admits large quantities of properly tempered light into the hall. In this connection it may be well to say that it is not pleasant to walk from the light front end of a hall into the dark back end. It is much more exhilarating to move into brighter light, and it is altogether more hospitable to present such an opportunity.

SOME HOUSE PLANS. — *Continued.*

CHAPTER XII.

A WIDE CENTRAL HALL OPEN AT EACH END. — LARGE ROOMS. — A PICTURESQUE STAIRWAY. — COLOR SCHEMES IN DECORATION. — DESCRIPTION OF FLOOR PLAN.

THE public likes a name. People like to be told that a building belongs to a certain style of architecture. Such a characterization simplifies matters, gives directness, and is therefore popular. The man who has wall paper to sell is pleased to say that he can decorate a room for his customer in the style of Louis XIV., or that of the Empire. The customer then knows what he has, is able to call it by name, and is therefore gratified. It often happens that in a relatively modest house there is found an Egyptian room, a Moorish room, one in the style of Louis XVI., and an Old Colonial room. This simple designation makes it easy for the owner to tell his friends and neighbors what he has. But, seriously, this question of style in buildings is often considered in a very honorable and proper way by people who are very much in earnest. They are anxious that all structures should be what they call pure in style; that is, if a building is designed in Italian Renaissance, that everything in connection with it should be exactly what the Italian architect would have made it at a given time. Many very precise and scholarly buildings are constructed in this way.

But in travelling through the European countries, one finds comparatively few of the earlier buildings absolutely pure in style; that is, structures which were begun with a definite idea in view, finished according to that plan, and left in that condition. On the other hand, it is very common to find, among the earlier structures, those which were much changed and altered during their construction, and which have been repeatedly added to and altered since. The Cathedral of Tours was finished in the thirteenth century. In the fourteenth century the façade and several of the front bays were removed. The new façade was not finished until early in the seventeenth century. One finds the marks of the early fourteenth-century spirit, and changes in style by decades down to the early seventeenth century. There are Gothic door-ways and windows in the lower part of the structure, and above the architecture of the Renaissance with its classic forms. This sort of thing is common, to a greater or less degree, in nearly all early buildings. Some of the notable structures where this has not been done are not so worthy of admiration. This admixture of style has given vitality to these buildings when it otherwise might have been lacking. Where the style is pure,

FRONT ELEVATION
FIG. 94.

there is often a preciseness and a definiteness of expression that is not altogether interesting. The lesson to be drawn from this is that our successful architecture will not come from the accurate following of certain definite styles, but from a knowledge of how and why all these things are done, combined with a spirit of independence and an artistic perception of the demands of our own time.

In this building one will find many details and forms which remind him of the style of Francis I. (Figs. 94 and 95). There is much in the details of this composition which belongs to that period of the Renaissance. On the other hand, the spirit of the work of the time of Francis I. could not be wholly carried through this structure.

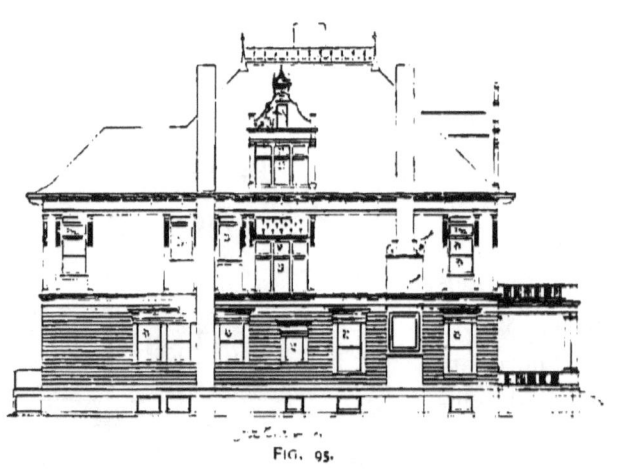

FIG. 95.

It would be quite absurd to undertake absolutely to transport such details into a wooden building. The construction of a window or a door frame in wood lends itself in a rather indirect way to the decorative forms which belong to that style. One who is honestly striving to do good work does not care to force a matter of this kind. Nevertheless, one finds in the Francis I. façade at Blois, in certain features in the Château at Chenonceau, and some of the buildings at Orléans, work in this style constructed in stone which must unquestionably have been suggested by wood details. A great deal which in the original is constructed in stone is quite as well adapted to wood.

One who studies this style feels the same justification in the liberal use of these motives that he does in the use of the more severely classic motives in the Old Colonial structures. The Old Colonial decorative forms were in stone, yet no one can question the great success which has attended their use in wood. It has been done with a candor, an honesty and freedom, which have brought their own reward. The Francis I. work is somewhat more decorative, certainly more exuberant in its method, and well suited to properly located American homes. But it is well to understand in connection with its use in this instance that there is no pretence of great fidelity. The forms of the mouldings, the outlines of the dormers, to some extent the disposition of the pilasters, the placing of the first-story string course, and the general treatment of the first-story walls are Francis I. in character.

FIG. 96.

On the other hand, the porch is somewhat more severe in its treatment than is usually found in the architecture of that time. However, the building is undeniably wood in its construction, finish, and decoration. It carries the stamp of a frame house.

A number of details of the exterior are given, in order

that those interested may see that the construction is good as wood construction, independent of its relation to any style. On the whole, these details are much less elaborate than those frequently used in the architecture of Francis I. In the adaptation of this style to this house there was primarily the limitation in cost, and secondarily the difficulty of securing the proper execution of rich detail. One frequently has to restrain himself because of the impossibility of securing artists capable of executing elaborate details. One must be contented ordinarily with mere form independent of the decorations which were associated with the original. It is much better, however, to have a perfectly fair surface than one covered with badly executed forms. It must be the mission of the art schools to train young men and women capable of doing a high grade of artistic decorative work. Until that time we shall have to content ourselves with the execution of simple outlines.

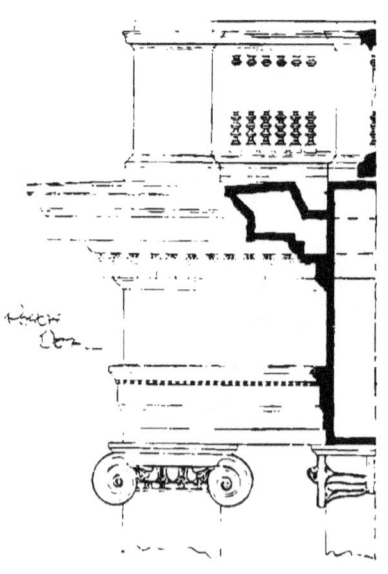

Fig. 97.

In a structure of this kind, where a good deal is made of the roof, where the dormers, chimneys, ridges, and decks are handled in a decorative way, one can afford more of repose in the lower part. If there is the opportunity of decorating the lower parts, it is well that it be done in a mild, sober spirit, in order to develop what is essentially interesting in itself. This idea is carried out in a manner almost brutal at the Château of Chambord. The lower part of the building is almost gross in its handling, while everything above the upper cornice line is brilliant in outline and highly decorative in detail. If the same spirit had been followed in the lower part the building would have been lacking in repose, which is happily not true in this great structure as we now find it.

Fig. 96 is an elevation of one of the dormers. Fig. 97 is a detail of the front porch. There can be no question but that a careful consideration of all the little details, the mouldings and the decorative forms, is absolutely essential to artistic success. The architect must give them the same quiet, thoughtful attention that a capable artist gives his picture. If this is not done, the material and labor which go into the building are misplaced — wasted.

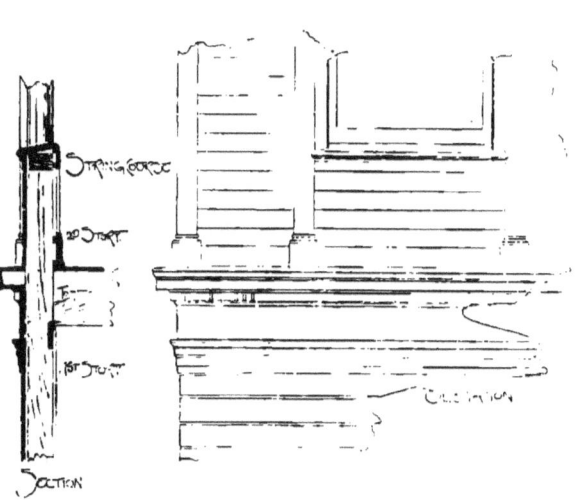

Fig. 98.—First-Story String Course.

This house, like all others, was planned without a very definite regard to the exterior. There always is, of course, an unconscious thought in the arrangement of a floor plan which measurably directs one, so that there is no great difficulty in putting the exterior into presentable form. This building (Figs. 100 and 101) has a centre hall fifteen feet wide, which extends from front to rear of the building. At the rear of the hall on the stair landing there is a properly divided opening which occupies the full width of the hall. Practically it is one window with five divisions.

The stairway (Fig. 102) gallery and the window which

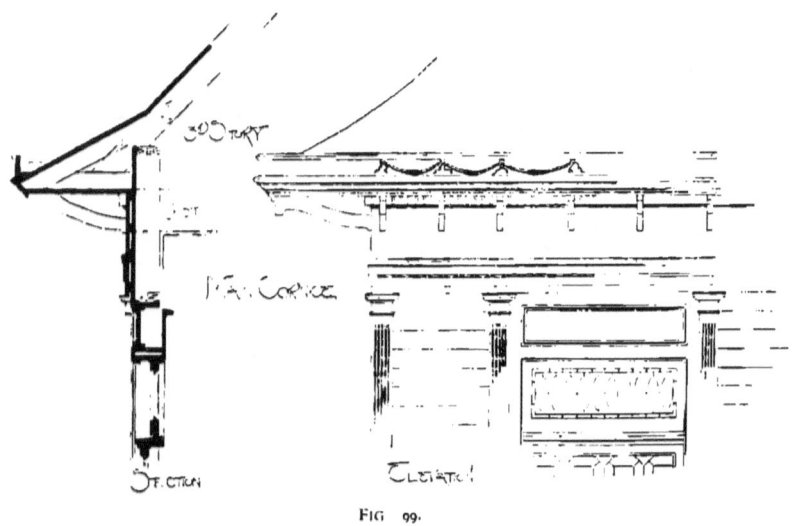

FIG 99.

is behind it are the central features in this composition. The great hall would not be interesting without something of this kind handled in a decorative way. Contributing to the general effect is a large mantel with fireplace (Fig. 103). To the front are the vestibule and the small screen-room, where one places wraps, overshoes, etc. With this hall thus emphasized, one has a basis for an interior composition. The long parlor on the one side, the sitting and dining rooms on the other, group well around the central feature. In arranging a plan of this character, the question is not altogether one of convenience, but is as well one

Fig. 100 — First Story.   Fig. 101. — Second Story.

of artistic composition. Not only should the convenient disposition of the halls and rooms be considered, but their arrangement in a satisfactory manner when viewed solely from an artistic standpoint

Standing in the middle of the hall after one passes the front door, the most interesting view is toward the fireplace, the stairway, gallery, and great window beyond. There are subsidiary pictures on either side. One is the sitting-room, with its mantel, windows, decorative wall-surfaces, and furniture. The other is the parlor, with its fittings and furnishings. Looking beyond and to one side is the dining-room, which is naturally a very brilliant composition. There is nothing more beautiful than such a room, with its great table and stately chairs, its sideboard next to the large wall-surfaces, and the large fireplace at one end. One can do nothing to a dining-room of fair size to take away from this stately quality. The mere placing of the table and chairs in the centre of the room, and the naturally geometrical setting of a table with its china and silver and glass, give a dignity and a preciseness which, with its natural beauty of color, nearly always make a dining-room one of the most artistic compositions about a house. This dining-room is finished in mahogany stained a rather dark red. The walls are almost a sage green; the ceiling a rather lighter green, enlivened with decorative figures in silver. The wainscoting is about five feet high. Little is made of the sideboard architecturally. It is more like a part of the wainscoting projecting into the room, and forming a kind of enclosed table. There are the shelves above, and plain surfaces to which one may attach plates or racks whereon may be placed glass or silver ware. There is nothing so nice about a sideboard

FIG. 102. — STAIRWAY.

as what goes on it, the glass, the china, and the silver, and it is fitting that the architectural part should not compete with what is intended to render it decorative.

The main hall has an oak wainscoting. Its wall coloring is tan. The decorations of the sitting-room and parlor on either side are in lighter shades of the same color, the parlor being rather more brilliant than the other rooms.

The connection between the kitchen and dining-room in this house is underneath the gallery of the stairway. From the china-room one may go to the cellar, or to a vestibule which leads to the rear porch. This vestibule opens into the kitchen as well as into the china-room.

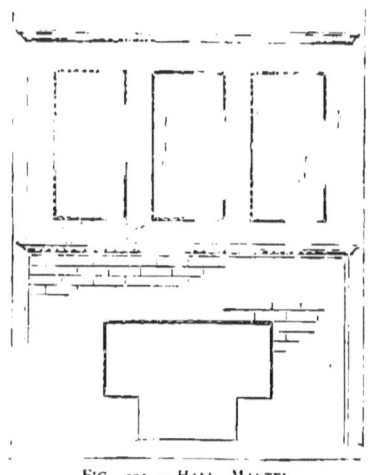

FIG. 103. — HALL MANTEL.

There is a driveway back of the house. The vestibule and china-room form a convenient passage to the front hall. It is intended that the passageway to the cellar under the main stairway shall be principally used in going to the billiard-room, which is under the dining-room. The arrangement of the china-room itself, with its glass-faced cupboard, sink, and tables, is clearly indicated in the plan.

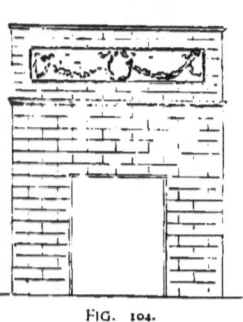

FIG. 104.

The second floor (Fig. 101) has a fireplace in the hall over the one below; in the principal bedrooms the grates of the lower floor are repeated. Two mantels are pictured in Figs. 104 and 105. Some of the closets are like those described on page 139. The bathroom connected with the independent water-closet, the store-closet, and other features are developed with the highest regard for detail.

The side hall on the south affords a direct passage to the attic without entering the main hall. In the attic there are two bedrooms, a large room occupying nearly one-half of the floor surface, and a store and tank room over the bath-room and water-closet.

The finish of the doors and windows is properly designated in Fig. 106. The panels of the doors are large and entirely flat and plain. The wood is oak, birch, or mahogany, either finished naturally or slightly stained. In either case the grain

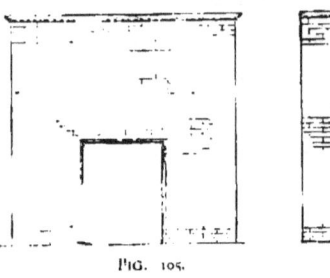

FIG. 105.

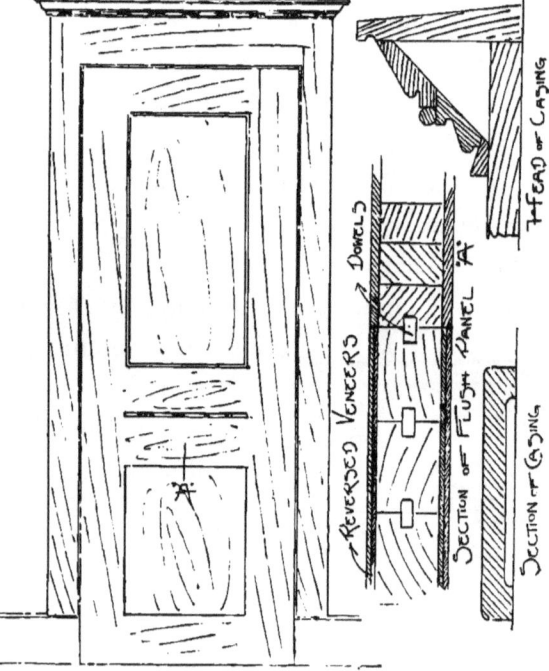

FIG. 106. — DOOR AND DETAILS.

of the wood is relied upon for the beauty of the finish. There are no mouldings or other disturbances of the plain surfaces. The lower panels of the doors are flush; that is, there is no recess. The entire face of this part of the door is perfectly smooth. Though a line is shown, as indicated by the drawing, it is merely the joints in the veneers and not a break in the surface. The woodwork under these panels has to be constructed with the greatest care. The details thereof are rather too complex for description in a general work of this kind. It may be well to say that the inner core of the panels of all of the doors is of soft wood; that under the broad surfaces this core is laminated and dowelled, and under the smaller surfaces it is merely laminated. Work of this character has to be done with the nicest skill, and by workmen of integrity.

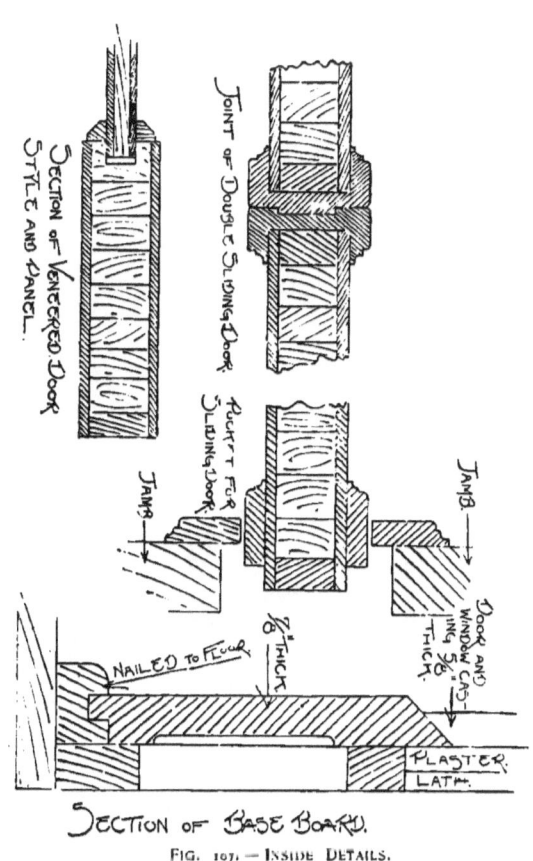

FIG. 107.—INSIDE DETAILS.

Fig. 107 shows a very acceptable way of bringing sliding doors together. With such an arrangement one can use the flat-faced hardware, which is altogether more interesting and satisfactory than the ugly astragal fronts.

Some of the mantels and other details are properly designated and scattered through this text. The general character of the finish is very simple. Nicely selected wood was used, and beauty secured in this way rather than by obliterating plain surfaces by mouldings, carvings, and other features.

SOME HOUSE PLANS. — Continued.

CHAPTER XIII.

A FINE LOCATION. — A RIVER FRONT. — PICTURESQUE STAIR-HALL. — A SMOKING-ROOM UNDER THE BALCONY. — WOOD CEILING. — DINING-ROOM FINISH. — KITCHEN DETAILS. — DOORS AND CASINGS. — GREEK PROFILES. — THE LOCATION PLAN.

A CONVENIENT floor-plan and an artistic exterior can be made for any location. It is manifest, however, that a given plan will not suit every location. The conditions influencing the planning of this house (Figs. 108, 109, 110, and 111) were all favorable, although somewhat unusual. The road shown on the location plan (Fig. 112) is about eighty feet above the bed of a river. There is a sharp bluff from the road to the river, and to the house a gentle slope. The frontage of the lot is about two hundred feet, and the depth about three hundred. It was

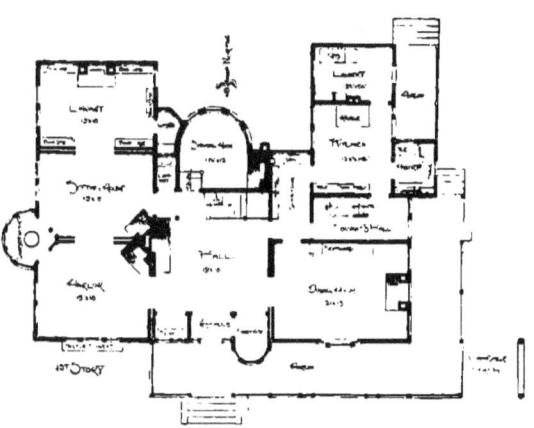

FIG. 108. — FIRST-FLOOR PLAN.

manifestly important to have the principal rooms of the house to the front. The sitting-room, the hall, and the dining-room of the first story occupy that position. Back of the sitting-room is a library, and back of the dining-room the side hall, china-room, and kitchen. The main hall extends the full depth of the house, with certain limitations. The lot and the general situation suggest this arrangement.

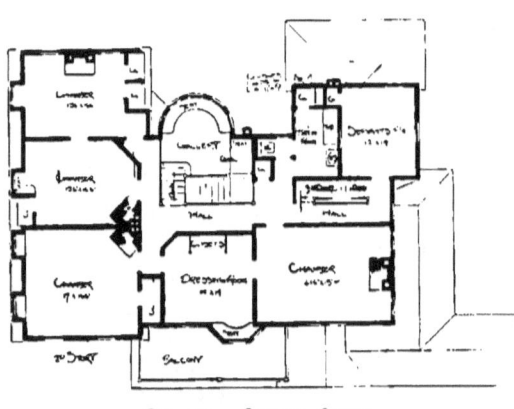

FIG. 109.— SECOND STORY.

One may enter this house from the front porch, the porch on the west, or the kitchen porch. Joining the sitting-room and library is a conservatory with glazed openings. In the centre of the conservatory there is a fountain.

As one enters the main hall there is a little cloak-room to the right. This can be entered without going into the hall proper.

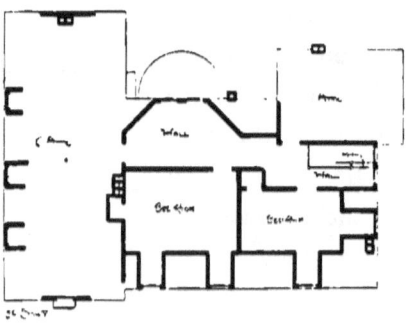

FIG. 110.— THIRD-FLOOR PLAN.

It can be measurably closed off from it by drapery. Thus the annoyance of having to place wraps and over-shoes in the hall, or of having to cross the hall to reach a place therefor, is entirely obviated. It is well-nigh impossible to keep a floor in good condition if one must pass over it with wet rubbers. This arrangement entirely does away with such a necessity, and makes a room of this hall. As one steps in the door one sees the large gallery which forms the stair-landing at the end of the hall. The stair goes up a little over half-way, and turns to a landing ten feet square, with a semicircular end and a half-domed ceiling at the top. In this semicircular end are five large windows of carefully designed

FIG. 111. — EXTERIOR.

stained glass. Under the gallery there is a smoking-room, which is reached by descending four steps from the main floor of the hall. In one end of this room, in a little nook, is a broad fireplace. The railing of the stairway continues around the gallery on the second floor, and thus adds to the general interest of the entire arrangement. Off from the hall, near the smoking-room door, is a lavatory. On the left-hand side is a large fireplace, with an eight-foot breast, the design of which is given in Fig. 113. From this hall there is an interesting view of the sitting-room and conservatory on one side, and of the dining-room, with its mantel and sideboard (Fig. 114), on the other side. Fig. 115 is an end of the library. In the front of the hall, looking to the porch, is a little nook with seats.

Another attractive outlook is from the sitting-room through the library to the fireplace with its window above. The dining-room is of mahogany, the hall of white oak, the sitting-room and library of birch stained a light red. The hall and dining-room have a wainscoting five feet high. The ceiling of the hall is entirely of wood, with large transverse beams and small lateral intersecting beams. The filling between is in plain surfaces. The outline view of the stair (Fig. 116) gives the stair panelling, the balustrades, and a general view of the gallery in the dining-room. One may go to the kitchen through the butler's pantry, in which is

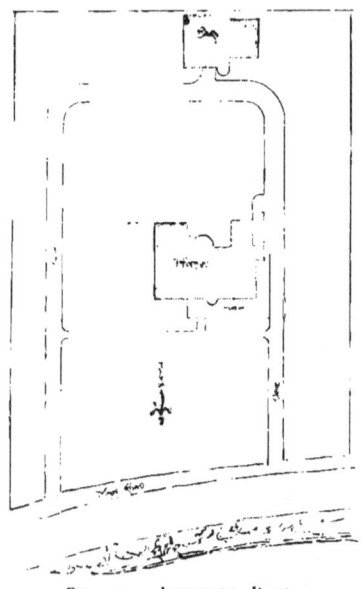

FIG. 112. — LOCATION PLAN.

FIG. 113. — HALL MANTEL.

placed a long cupboard, with glass doors in its upper section, and panels and drawers in its lower part. At the end of this room under the window is the sink for washing

FIG. 114. — DINING-ROOM MANTEL.

the china. The connecting doors between the butler's pantry and the dining-room and the kitchen are on double spring hinges. The kitchen has cross ventilation; that is, there are openings on each side. At one end is a tile hearth, on which is placed the range; at the other end, the tables and sinks. The detail drawing shows the general construction of this end of the kitchen. On the wall back of the tables there is a lining of slate (Fig. 117). The tables, which are portable, are set against the slate, and extend over the sink. The drawing shows this. Near this end of the room is the pantry, with its cupboards, tables, drawers, flour bin, and wall hooks for utensils. At one side a refrigerator is built in, with a door at the back, through which ice may be placed from the porch. Back of the kitchen is the laundry with its tubs. In this room is a closet connecting with the clothes-chute in the bath-room above. From the side hall one may go to the cellar, and to the second and third floors.

The room over the dining-room and the one over the

FIG. 115. — END OF LIBRARY.

porch are arranged in one suite. The closets of the dressing-room are of the type described on page 139. The bath and water-closet room are independent, as has been shown in others of these plans. The walls of both of these rooms are lined with marble, and the floors are covered with tiling. A linen-closet is shown next the water-closet entrance, and another store-closet off from the bath-room. The doorway from the main hall to the side stair-hall renders the latter and the bedroom connecting with it separate apartments.

On the third floor the space over chamber and dressing-room is given to two bedrooms. Over the bath and bedroom next to it is a water-tank. The space over the two west chambers is plastered and used for storage.

As to the plumbing, it may be well to say that the water-closets are of the siphon pattern, that for the servant being located in the basement. The washstands are fourteen by seventeen inches over bowls, with secret waste and overflow, the slabs supported on nickel legs. The bath-tub is roll rim, porcelain-lined, with modern fixtures. The china sink is porcelain with slate back; the kitchen sink iron, porcelain lined; the laundry tubs brown earthenware, with brass trimmings and connections. All other brass-work throughout the house is nickel plated. The house is lighted by electricity and gas. Natural gas is used for heating water, for the heating apparatus in the basement, and grates in the various rooms.

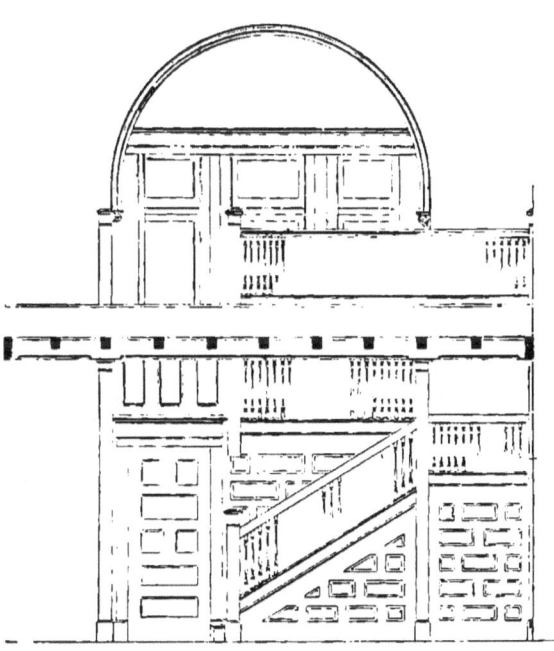

FIG. 116 — STAIRWAY.

Fig. 118 shows a door and casing. A is a section of the head of the casing. The little incision, which is designated by the arrow B, is for the purpose of giving a sharp, clear, though delicate, line at this point. The mouldings are very fine and simple. This incision is to be observed elsewhere in connection with this casing-head, and is used for the purpose of accentuating the fine lines desirable in connection with these profiles. The mouldings are thus sufficiently conspicuous without being coarse and heavy. These and all of the other mouldings used are of the Greek pattern rather than the Roman, which is more commonly used with Old Colonial outlines. This same thought obtains in the designing of the mouldings of stairway, mantels, string courses, and other internal and external details.

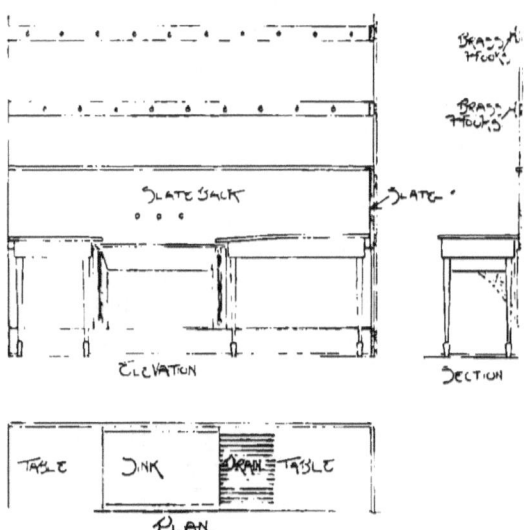

FIG. 117.—KITCHEN SINK AND TABLES.

The size and location of the lot suggested the general floor-plan. The floor plan and the view from various points suggested the external outline. The details were developed with the idea of making a near view pleasing.

The location plan indicates the disposition of the

grounds. The space immediately in front of the house from the porch walk to the road is left entirely plain. It is not crossed by walks, or disturbed by flowerbeds or other features. This stretch of plain lawn, extending somewhat more than the full width of the building, gives a better setting to the structure than if there were anything immediately in front to distract the eye or to divide the interest. There is a drive on the east side of the house, and a walk on the west side. The drive continues to the stable in the manner shown. The house has a very good setting of trees some distance in the rear.

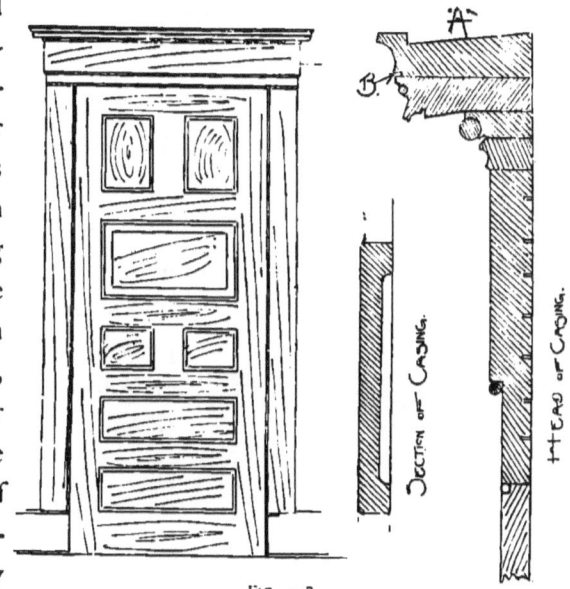

FIG. 118.

SOME HOUSE PLANS. — *Continued.*

CHAPTER XIV.

A STABLE. — A WATER-TOWER. — PUMPS — A FLOOR PLAN. — INTERIOR DETAILS. — DRAINAGE OF THE FLOOR AND STALLS. — AN ODORLESS MANURE-BIN. — STALL CONSTRUCTION. — THE MAN'S ROOM.

THE stable shown on the location plan in the previous chapter contains some features worthy of description (Figs. 119 and 120). The tower, which shows so conspicuously in the elevations, contains a tank for well-water, which is used for various purposes in the house and stable, and for sprinkling the lawns. In this instance the water is supplied by a pump, the power for which is the direct pressure of natural gas, which in this locality happens to be thirty-five pounds per square inch. Such power is only available in restricted sections. In other sections water motors, gas, naphtha, gasoline or hot-air engines, electricity or windmills, may be used. Sufficient to say, a water plant of this kind is everywhere available. There is only required some engineering ability

FIG. 119. — BARN, FROM REAR.

in planning it The operation of any such apparatus is exceedingly simple, needing only a little faithfulness in oiling and the courage to refrain from tinkering with it.

The plan (Figs. 121 and 122) shows the carriage-room, sufficient in size for the storage of four vehicles, and a stall-room, containing two box stalls and one ordinary stall. Connected with both the stall and carriage rooms is a large harness-room. Out of this room is a stairway to the second floor. This upper part is used for a man's room and storage. In another structure built from about this plan the large room was used as a gymnasium and playroom for boys. One of the closets could be arranged as a dark room for amateur photography. The space over the stall-room is used exclusively for hay and feed storage. The dormers on the east side of the hayloft afford passages from which to throw hay into the stall-racks below.

FIG. 120 — BARN, FROM FRONT.

There are little doors hinged into the floor, which may be lifted, and, as the dormers are sheathed down to the floor as far back as the hinge of the door, the opening of this door forms a continuation of the hay-chute up into the loft. The door may be hooked open to the side of the dormer, and held in that position at the will of the attendant. There are large doors at the north and south ends of the loft. The rooms over the carriage-house are plastered with hard plaster. The walls which separate the hay-mow from these rooms are sheathed on the hall side with dovetailed sheathing, which forms a surface which can be plastered on one side, and a beaded flooring-surface on the other. This is not the ordinary dovetailed or Byrkit lath, as it is called, but is a high grade of dressed yellow-pine flooring, which has the lath cuts made on the side to be plastered.

All of the framing-stuff, sheathing, and exposed lumber on the interior of the structure is of Southern yellow-pine. In the section where this stable was built the cost of this material is about the same as the Northern pine. It has the advantage of being free from knots and other blemishes. The exterior woodwork is poplar, excepting where shingles are used, and they are pine or cypress.

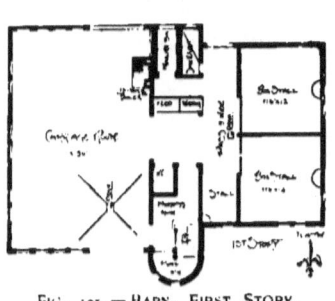

FIG. 121. — BARN, FIRST STORY.

The stall-room, as stated, contains two box stalls and one ordinary stall. The box stalls are twelve feet square. In case of emergency one or both of these stalls could be provided with temporary or permanent partitions, and the capacity thus be increased from three to five horses. Even then the stalls would be larger than those in common use. The water-closet in the room off from the stall-room is of the type known as out-door closet; that is, the valve and trap are well underneath the ground, so that they will not freeze in winter. This closet connects with a sewer for drainage, and with the tank service for supply. The feed-mixing box and watering-trough are at the side of the passage, between the stall-room and the carriage-room. Connecting with the outside of the stable on the north is a sawdust bin. While straw probably makes a better bedding-material, sawdust is largely used in many sections. This bin is hoppered in the direction of the inside opening.

The arrangement of the manure-bin is worthy of special attention. It is the lack of care of this material which so often renders a stable an obnoxious place. This manure-bin is constructed with vitreous salt glazed brick laid in Portland cement. The floor thereof is of Portland-cement concrete to the depth of four inches, and slants to the outside. The bin is arched over at the top with the material previously specified. On the inside of the stable there is a close-fitting cast-iron door set four feet from the floor. On the outside there is a wrought-iron door four feet high. This door is the full width of the bin. At the top it is perforated with a number of small openings. This permits the passage of air into the bin,

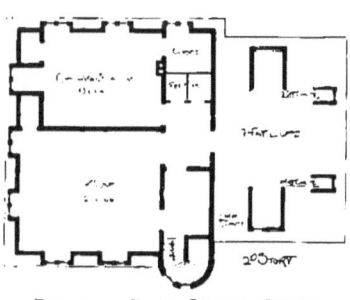

FIG. 122.—BARN, SECOND STORY.

and thence to a special vent-flue which connects with it. The movement of the air in this flue is accelerated in cold weather by the heat of the grate which heats the carriage-room, and by a special natural-gas burner in the vent-flue itself during other times. Thus it is impossible that there should be any odor from the storage of manure. In foreign cities one finds stables opening into the court-yards of splendid establishments, and with ordinary care, even without the precautions which are made possible in this instance, there is never any odor.

A very large proportion of the houses in the Faubourg Saint Germain, in Paris, are provided with driveways into the courts, which are relatively small in size, yet from which the doors open into the stall and carriage rooms. There is no odor to remind one of their presence. Certainly there need be none in a stable of the character we are considering.

The stalls on the wall side are lined to the height of six feet and a half with wainscoting. The other walls and partitions are formed with two thicknesses of seven-eighths-inch flooring nailed together into reverse diagonals and capped on the top with special cast-iron covering. These walls and partitions are four feet high. On these are placed wire screens two feet and a half high. The windows which light and ventilate the stalls are above the wainscoting and hinged at the top. They are opened by cords, which pass over pulleys to the west side of the stall-room. The cord is fastened to catches at the bottom of the sash, so that the fastening is released by pulling, and the sash opened.

The stall floors are formed by first laying a four-inch bed of concrete. On this is spread a light flow of pitch, and over this two coats of roofing-felt stuck with pitch. In the concrete are bedded four-inch by four-inch sills. On these is laid secret-nailed, two-inch tongued and grooved pine flooring. This floor has a slant from a corner to the catch-basin shown on the outside of the stall. This slant, of course, is very slight. There is some difference of opinion as regards the best kind of a stall floor. The advantage of this one is that it is water-tight, has good drainage, and is capable, if well bedded, of being kept absolutely clean and pure. Clay floors, which are favored by many, are absorbent, and cannot be maintained in a condition to afford good drainage. The floors of the entire stable outside of the stalls are of vitrified paving-brick. There is first laid a bed of sand, followed by a smooth bed of Portland-cement concrete, in which vitrified paving-brick are laid on edge and slushed with liquid cement. The floors in both the stall and carriage rooms slant to catch-basins, which are connected with the sewer. Thus a hose can be turned on the floor in either room, and it will thoroughly drain itself. The slant of the floor need not be enough to be noticeable in walking over it. There are hot and cold water connections in the carriage-room. The hot water is provided by a forty-five-gallon plumber's boiler, which is connected with a coil heater, either in the grate or above a natural-gas burner.

The man's room on the second floor is heated by a register connection with the hot-air flue, which joins a wrought-iron brick heater in the carriage-room. By this arrangement fresh air is drawn in from the outside of the building on the west side, through a six-inch vitrified pipe back of the wrought-iron lining of the grate. In its passage into the hot-air chambers it is heated by the grate-lining and the smoke-flue, which are independent and set inside of the hot-air flue. Thus the waste heat of the grate is used in heating outside air which comes through the vitrified pipes, the flue, and into the man's room above.

SOME HOUSE PLANS. — *Continued.*

CHAPTER XV.

THE HOUSEKEEPER AND THE FLOOR PLAN. — THE SOBER-MINDED CLIENT. — THE ONE WITH PICTURESQUE TASTES. — SOUTH GERMAN ARCHITECTURE.

IN building a house the floor plan should receive first consideration. Every housekeeper has certain ideas of arrangement; and while it is true that the housekeeper may know very little about the principles of house-planning, the arrangement of rooms, halls, stairs, closets, etc., she is nearly always able so to express herself that she practically controls the general arrangement of the plan. Many details may have to be changed; yet, as in the house shown in Figs. 124 and 125, certain general requirements affect the entire arrangement. A large hall and parlor were demanded for the front on the first floor; back of the parlor, the library; in the rear of the hall, the dining-room. A rear hall connecting with both library and dining-room was desired. The demand for the second floor was that there should be five bedrooms, and that the family room over the parlor should connect with the hall bedroom and the one over the library. The guest-room was over the dining-room, and the servant's room over the kitchen. The rear stairway continued to the attic. The bath-room and water-closet room were separate; the laundry was provided in the basement. The arrangement of all details, such as the closets, kitchen, pantries, and other matters of like character, was left entirely to the architect.

Clients usually do not say a great deal to an architect in whom they have confidence, in regard to the exterior. He knows the limit of expenditure, the general taste of the owner, and governs himself accordingly. One client may demand a very quiet, simple exterior. That is, the architect may feel that such a design would be best suited to his taste. Yet another client may be more picturesque in his tastes, rather more brilliant in character, and the architect who understands his client will make a design suited to his character. There are others who wish what is ponderous or dignified, others what is chic or quaint. Thus there is developed a perfectly natural and proper demand for variety in external treatment. If all of these clients be people

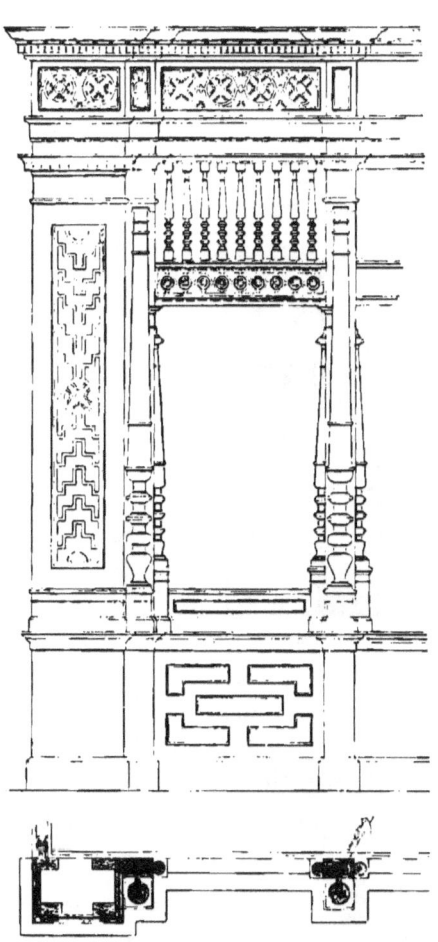

FIG. 123

of correct taste and refined manners, their varying demands would be properly considered by a skilful architect.

In connection with a simple plan of the kind here given, it is interesting to show at least two methods of treating the exterior. The first one (Fig. 126) is very simple and quiet, yet having every bit of decoration carefully, thoroughly, and seriously developed. Each moulding on the exterior has been a subject of careful thought, each bit of decoration worked out in a way to harmonize with the relatively severe treatment of the other details. The elevation of this structure shows the spirit in which this work was carried out. Such a design might exactly suit the demands of one client.

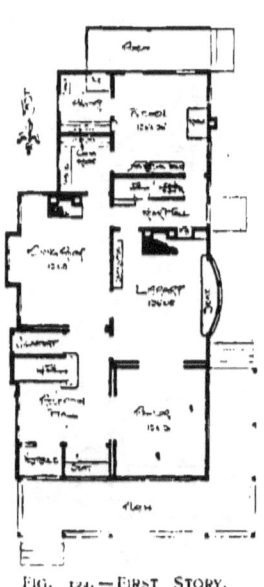

FIG. 124.—FIRST STORY.

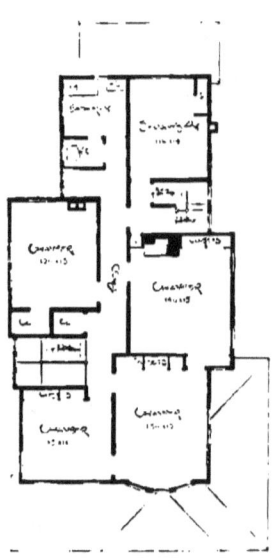

FIG. 125.—SECOND STORY.

Take another instance. The architect notices that his client is rather picturesque in his tastes; that a coarse, extravagant design might please, if presented to him; that such a design would be selected rather than one simpler and more refined, not because of inherent coarseness, but because few people are well enough acquainted with the details of design to make fine distinctions. While they wish a picturesque structure, they will be more than pleased if its picturesqueness can be obtained in a refined, studious manner. These people will appreciate carefully considered detail quite as much as the other client. In all probability they would not be able to judge as to the exact character of the details until they were executed.

FIG. 126. — FRONT ELEVATION.

If the public demand for rather brilliant structures, which has been undeniable in recent years, had been met by architects who could design all of the details in a refined spirit, we should have wonderfully picturesque cities rather than a large number of coarse and common ones. The house of the client who wishes a large number of gables, a tower, dormers, balconies, etc., is nearly always a violently ugly one. This is true because the details of the structure are bad. Such a design is no more complicated than many which we see in the pictures of the

FIG. 127. — FRONT ELEVATION.

Breton, English, and German architecture. In each of the countries named we find beautiful picturesque streets. The work is done by artists. With us it is quite different. As has been said before, the client wishes the picturesque. The architect is able to give him only what is confused in outline and coarse and common in detail.

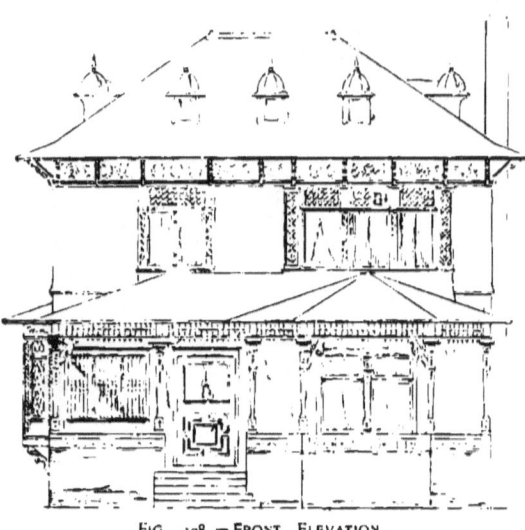

FIG. 128 — FRONT ELEVATION.

In the exteriors which are given of this house, the first has mouldings taken from the Roman architecture. These are particularly well-fitted for execution in wood. This harmonizes with the other design, and is capable of being carried out at a moderate cost.

The more picturesque design (Fig. 127) is in-

FIG. 129.—FIRST STORY.

FIG. 130.—SECOND STORY.

spired by the South German half-timber architecture. Nothing could be more picturesque, and yet it is highly refined in every detail. The filling in between the timbers is of concrete. The drawings tell their own story.

The elevation (Fig. 128) and two floor-plans (Figs. 129 and 130) are of a house not greatly different from these. An especial description is not necessary. The stairway of the house is shown in Figs. 131 and 132.

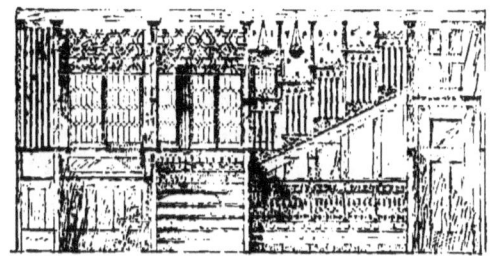

FIG. 131.—STAIRWAY.

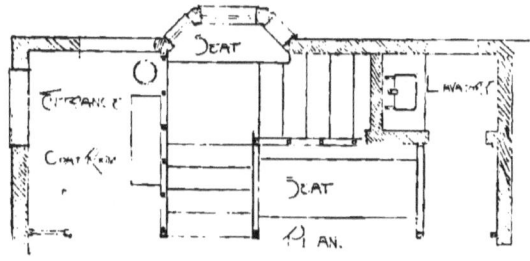

FIG. 132.—PLAN OF STAIRWAY.

SOME HOUSE PLANS. — *Continued.*

CHAPTER XVI.

IN A FOREST. — SENTIMENT WHICH CONTROLS THE EXTERIOR. — A WIDE HALL. — THE MUSIC-ROOM. — THE END OF THE DINING-ROOM. — THE SIDEBOARD. — FLUSH-PANEL DOORS. — A SCREEN. — NATURAL-WOOD FINISH FOR THE EXTERIOR. — ROMAN BRICK. — BYZANTINE DETAIL.

THIS house (Figs. 133 and 134) is situated in a large grove of forest trees, near the bank of a winding river. The purpose in its planning was to get the principal rooms to the front; therefore it is a broad and relatively shallow structure. The first idea as to the exterior was that it should be perfectly simple and natural in general outline. Therefore, one finds no gables, towers, or other detail to bring one part into greater prominence than another.

FIG. 133. — FRONT ELEVATION.

There was the feeling that in this great grove a structure simple in outline and fine and carefully considered in detail, would be reposeful in its association with the picturesque surroundings.

The sentiment which controls the exterior dominates the interior arrangement (Figs. 135 and 136). In a perfectly natural way there are developed a number of cosey corners and special features which have been disassociated for the most part with our house-building during the recent past. Large plate-glass show windows staring at the street do not bring the cosiness and the homely comfort which belong to quiet home-building.

FIG. 134.—SIDE ELEVATION.

This plan is formed upon broad lines. The wide hall and the long rooms on either side form the living part of the house on the first floor. But these rooms are large, and so divided as to obliterate the idea of mere roominess. The separation of the little cloak-room from the hall is more complete than in any of the houses which have been shown. Beyond the vestibule is an ample window-seat (Fig. 137), which is separated in a decorative way from the hall, at the same time that the entrance to it is direct, and the view through its windows to the front unobstructed. There is, moreover, a pronounced decorative separation between this recess and the large hall. The view from the hall itself is not intended to be startling. There are quiet details and features of interest in the library, dining-room, and music-room beyond the hall, and in the stairway. The purpose of the designer was to keep everything quiet and simple. While there is a wainscoting five feet in height encircling hall, music-room, and dining-room, the surfaces are perfectly plain. There are no panels in either the wainscoting or doors. The broad surfaces of woodwork are covered with selected veneer of quartered oak, and are finished with a dull surface.

FIG. 135.—FIRST STORY.

The music-room in the rear is three steps higher than the hall. In it there is a little alcove in which can be placed an upright piano. This alcove is not enclosed on the side toward the hall, but is merely screened off from it by decorative woodwork. This is quite open, and would not interrupt either view or sound. At the back of the music-room is a broad seat under a window, which is quite high above the floor (Fig. 138). It is to be noticed that there is a slight separation between this little music-room and the main hall, in the form of a screen (Fig. 139). As one stands near the front entrance to the main hall one sees the large fireplace built of light-colored tiles, the elevated music-room, the screens which separate it from the hall, the window-

FIG. 136.—SECOND STORY.

seat in the rear, and the light coming from a broad window on the main stair-landing above. The source of this light is not seen from this point, but one feels the light, warm tints from the broad window above. The decoration of the room and the treatment of the woodwork are rich and rather sombre, and in walking

FIG. 137.—LOOKING INTO ALCOVE FROM HALL.

towards the stair-hall there is the impression of the brightness that is beyond. There is a buoyancy in an effect of this character which cannot be explained. One cannot analyze it, but one must feel it when in the room.

The end of the dining-room (Fig. 140) which is seen from the hall is divided by a beam extending from wall to wall, and supported on either side by a pilaster and column. The mantel is very broad — seven feet. The treatment of the tile around the hob-grate is perfectly simple. The colors are gray. Above is a plain mantel-shelf, and yet above this a space for a picture. The space between the pilaster and the wall provides room for a window-seat, above which is a leaded-glass window. The space from the under side of the beam to the floor is only eight feet. Hence there is a distinct separation between this grate-hearth and window-seat space and the dining-room itself. This makes one of the cosey corners of this house. While it might seem that the cutting off of this end of the dining-room would reduce its seating capacity, it is not so, for the reason that the woodwork does not project beyond the tile-hearth, which is, in reality, the limit of the available space in a dining-room.

FIG. 138. — WINDOW-SEAT IN MUSIC-ROOM.

The west window of the dining-room is placed high from the floor. Thus, while it is wide it is not long, and being in three divisions does not cause a glare. The glass of this window is uncolored, and set in small geometrical patterns leaded. The sideboard is opposite the centre of this window. This sideboard, like the others which have been described, is architecturally simple, and depends upon the articles of china, glass, and silver for the decorative effect. A sideboard which is made up of pre-

FIG. 139.—SCREEN BETWEEN HALL AND MUSIC-ROOM.

tentious woodwork, with elaborate details, is out of place. The architecture of a sideboard cannot be expected to compete in a decorative way with the articles which it holds, and for that reason may well be kept quiet. The material may be never so rich, — oak, mahogany, birch, or other hard wood, — but it is an utterly misplaced effort to make it obtrusive through its own detail.

The library contains features of interest such as belong to such a room. At right angles to two of the bookcases are stationary seats, which are illustrated in the sketches in another part of this book. The aptness of this arrangement is at once apparent. There is a window-seat projecting from this room on the north. Through the north door the stairway is visible.

The doors (Fig. 141) have all of their panels flush with the rail, though the panels themselves are outlined by a rather delicate moulding. Their surfaces are of selected veneer, glued to a laminated core of poplar wood.

FIG. 140. — END OF DINING-ROOM.
(Painting built in over mantel.)

The details on page 197 show their construction and the precautions taken against warping. There are so many features connected with the interior of this structure that plain woodwork is necessary to prevent confusion.

The connection between the kitchen and the dining-room through the butler's pantry, and the fittings of the kitchen itself, are practically the same as those which have been described elsewhere. The laundry and servants' water-closet are in the basement.

On the second floor the bedrooms, halls, closets, and various fittings are properly indicated by the drawings. Two mantels on this floor are shown in Figs. 142, 143. The bath-room is separate from the water-closet room, and the general principles of the planning of the second floor are not essentially different from those which have been considered elsewhere. The floor of the bath-room and water-closet is of tile, and the walls to the height of six feet are of marble.

The exterior treatment is as simple as possible in its outline, though the details — that is, the cornice, mouldings, and window-caps, the porch columns, the dormers, string courses, projections, brackets, dentals, etc. — are all developed with the greatest care. It is necessary in a simple structure of this character to give much thought to all the details in order to prevent an appearance of commonness. This same building with crude mouldings and unstudied forms would be anything but interesting.

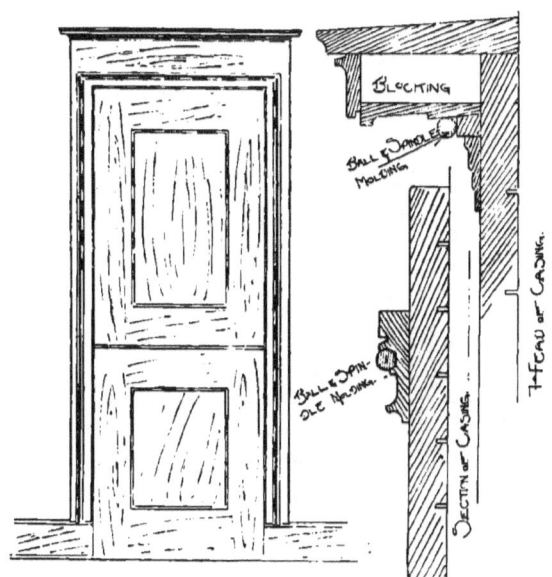

FIG. 141. — DOOR AND CASING.

It would be merely common. But with all these things carefully considered and executed the very simplicity of outline adds to its refinement and interest.

The foundation of this structure, which shows above the ground, and the exposed chimneys are all laid in Roman brick, which are about one and one-half inches thick and ten inches long. Their color is tan, mottled in blue. At a distance the effect is gray. The exterior of this structure is finished in natural wood; that is, the white pine was filled with shellac and then varnished. No paint is used on any part of the exterior. The writer thinks that this kind of finish must in time largely take the place of exterior painting. It is not expected that a waterproof varnish will retain its finish, nor is it intended that it should do so. The exposed parts of the wood will turn gray. Those under cover, such as the cornice and the parts under the porches, will turn to yellows and light reds, and blend with the grays in a very natural and interesting manner. In a structure situated as is this one, this coloring cannot but be attractive. The only houses which one sees in Europe which are interesting because of their paint are the Hildesheim structures. As we have said, the Swiss cottages are greatly enhanced in beauty through the weathering of the fine wood out of which they are made.

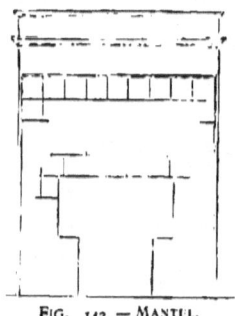

FIG. 142. — MANTEL.

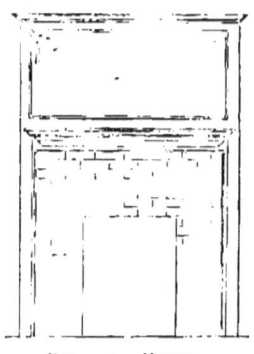

FIG. 143. — MANTEL.

On the exterior, under two of the windows, are flower-boxes. These boxes are of wood, lined with lead, and are provided with a drain which carries the drip-water to the ground. The boxes themselves rest on iron brackets. When made and supported in this way there is no chance of decay in any part. A few delicate vines trailing from these boxes, and the bright color of geraniums, are quite in character with the general conception of this building.

FIG. 141

SOME HOUSE PLANS. — *Continued.*

CHAPTER XVII.

A LARGE NUMBER OF ROOMS. — BRICK AND STONE. — RENAISSANCE FORMS. — PICTURESQUE ROOF. — THE GABLES.

THE plan (Fig. 145) given in this connection is a departure from the modern idea of consolidation in house-planning. There is a large central-hall, a separate stair-hall, with parlor, dining-room, and kitchen on one side of it, and sitting-room, library, office, toilet and closet rooms on the other side. No one will question the general benefit of simplifying a floor plan. We quite frequently see splendid houses which have only a large hall, a dining-room, library, and kitchen on the main floor. In this case the hall has its nook into which the formal caller may be ushered, and the hall itself, the library, or even the dining-room, may do service as a sitting-room. This has a ten-

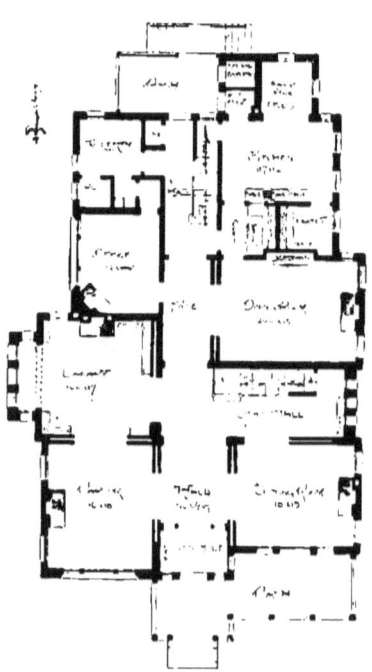

FIG. 145. — FIRST STORY.

dency to reduce complication in house-keeping. Much time is wasted through badly arranged floor plans, and a great deal of energy lost through the inexorable demands of over-housekeeping. If one were not acquainted with the conditions which dictated this floor plan, he might well class it among those which are too complicated. It may be well to state that the house was never built, though it was seriously considered. It was planned for a large lot, some two hundred feet front and three hundred feet deep; the ground was covered with magnificent trees, in the midst of which was the place designed for the house. The second floor, which is not given, contains six bedrooms, two dressing-rooms, bath-room, and water-closet; the attic floor, a large room in front and two bedrooms and a bath-room in the rear.

FIG. 146.—FRONT ELEVATION.

This house was to have been of brick (Figs. 146 and 147). Its great size presented certain difficulties in the composition of the exterior. If the north side had been continued through in a straight line in so far as it relates to the sitting-room, stair-hall, and dining-room, it would have been heavy, not alone because of the monotony of the wall surface, but on account of the depression of the roof above. As a relief from this, the stair-hall is recessed a little so that two gables are justified on this side of the house.

As shown by the elevation, this, together with the treatment of the large window over the stairway, relieves this side of the house from all danger of heaviness, which would result from its natural form and great size. The arrangement of the gables at the side improved the outlines of the front. The same difficulty of largeness was to be feared in the front; thus it had to be so developed as to enliven the structure without making it frivolous. In the front the surfaces are flatter than on the sides, and the mass more bulky. As a relief from this there was chosen a detail

FIG. 147. — SIDE ELEVATION.

somewhat decorative, at the same time that it preserved dignified outlines. Without adhering distinctly to any model, the general style of this structure is early Renaissance, and of a period which admits of some considerable decorative detail. Thus we have a sedate outline and the enlivening influence of rather brilliant decoration.

Where the details are sufficiently refined in character, the danger of over-richness is nearly reduced to the minimum. The roof-lines of this building, as shown by the drawings, end in a graceful curve at the bottom. This prevents a harshness which would result from running the roof-line straight down to the point of the cornice. The use of a large number of rather brilliant dormers in the roof prevents this part of the structure from being heavy. If one studies the detail of the wood-carving on the gables, he will see that it would take the hand of an artist to execute them. It was intended to have this woodwork carved by an artist, and placed in position along with the other work as the house went up. This is also true of the decorative woodwork of the porches and window-frames.

In order to carry the color of the body of the structure into the roof on the north side, the front dormer-walls are designed in brick, and are a continuation of the lower brick walls without the interruption of an east and west cornice-line at this point.

The base course from the top of the water-table is of irregular coursed masonry; otherwise the lower part of the structure contains a relatively small amount of stone, but as the building approaches the top it becomes richer in this material. This may be seen most readily in the front view, though it may be noticed on the side as well. This enlivens what would otherwise be a rather cumbersome and heavy structure. The character of the detail removes the impression of over-richness.

It is impossible to adapt the earlier styles to the requirements of an American home. There is no reason why it should be attempted. They were all developed through the practical working out of the problems of the people of their time, and as our habits and requirements are quite different from those of past ages, it is absurd to expect that one should preserve the traditions of an earlier style, and at the same time meet directly the requirements of our own time. All that one can do, and all that a conscientious artist wishes to do, is to preserve the spirit of the style in which he has undertaken to work. If one uses the style of any particular time, one can only develop it through the use of its decorative features. Everything else must belong to the spirit of the time in which he lives.

If one is designing a house of this kind in the style of the Renaissance, it is not to be expected that it will be clearly like any of the Renaissance monuments of the sixteenth and seventeenth centuries. This is distinctively an American dwelling, with as much of the spirit of that time as can be developed, and with certain details which belong to that period. It is a mistake of the modern German architects that they undertake exactly to reproduce motives, if not entire structures, of earlier periods. This is mere copying, and the result reflects discredit to all concerned. The buildings of Berlin and its suburbs have a factory-made look which far removes them from great works of art. They are unsatisfactory not merely because they are copies, but also because they are not beautiful. In art the end always justifies the means. One sees this sort of thing in all Germany. They, with their academic methods, their formal processes, naturally walk into this error. We Americans are far removed from any danger of this kind. We are greatly lacking in scholastic acquirements. If a little more academic knowledge could be infused into our architecture, it would lead us from much that is wild and irrational. We are at one extreme, and the Germans at the other.

SOME HOUSE PLANS. — *Continued.*

CHAPTER XVIII.

STONE HOUSES. — A HOUSE FOR A WIDE LOT. — A PLAN WITH LIVING-ROOMS TO THE SOUTH. — THE FRONT. — FIFTEENTH CENTURY.

THE stone house lends itself to city surroundings. It matters not if it have the conventional street-front, or if it be a detached structure. Even where a house is built connecting with others, there is an independence and a natural strength about stone which preserves the house as a thing apart. The mere repetition of a single design in houses destroys the idea of domesticity. A row of houses of the same general character cannot be associated with the home idea. Certainly there can be no personality in one of a great number of houses of this kind. It is easier to give houses each its own character when they are constructed in stone than when they are constructed in red brick. In a detached house, which is certainly the best place to display stone architecture, the independence, the separateness from surrounding buildings, is emphasized. It is a particularly happy circumstance when the yard can be large and the building surrounded by trees, and particularly great forest trees.

The introduction of Romanesque details in domestic architecture had not a little to do with bringing stone prominently into use in our dwellings. Mr. Richardson was responsible for this movement. He has had many imitators, but few who have worked with the same knowledge. The use of the forms of the old monastic and ecclesiastical architecture to clothe the houses of our own

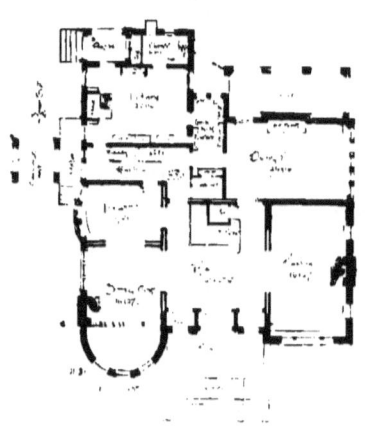

FIG. 148.—FIRST STORY.

times is, in itself, somewhat inconsistent, though Mr. Richardson carried out this style in a more refined way than belonged to it originally. His imitators, however, have not been so happy, for the reason that they have not been so thoughtful or intelligent. Their interpretation of the Romanesque is an architecture abounding in great semicircular arches, heavy lintels, ponderous dormers, and faced with large quarry-faced stones. This is often done without the compensating influence of finer, more delicate details.

A stone building is, in itself, strong in character. To preserve the relation to home life, its details should be fine and the general outline pleasing, rather than heavy and dignified. The French architecture of the late fifteenth and early sixteenth centuries lends itself most naturally to domestic purposes. Previous to this time the pretentious domestic architecture had been dis-

FIG. 149.—SECOND STORY.

tinctively military in purpose. While it was naturally picturesque, and while the details were fine, it had not assumed the distinctive social features which altogether changed the general conception of the earlier châteaux of that time. Those built during the fifteenth and sixteenth centuries were decidedly social in character, and while they were without the distinct home quality of our own time, the social element was conspicuous in the form and all of the details.

The floor plan, which is given in Figs. 148 and 149, is for a city house on a wide lot. There is a broad hall in the middle of the house. The stairway leads to a broad landing between the first and second floors. The parlor, dining-room, sitting-room, the little library, and the connection between the porte-cochère to the side hall, the kitchen, butler's pantry, and kitchen-pantry are arranged according to the gen-

FIG. 150. — FRONT ELEVATION.

eral principles which run through all of the plans of this book. The second floor suits the individual requirements of the people for whom it was planned.

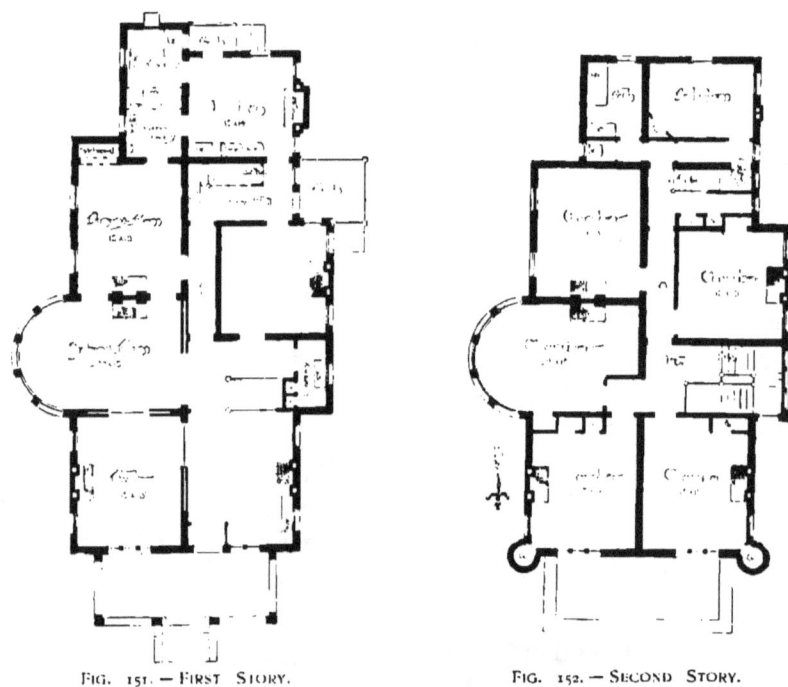

Fig. 151.—First Story.   Fig. 152.—Second Story.

While the controlling principles of house-planning may be always the same, there must yet be personal habits and circumstances which control the details. On the third floor of this building there is ample space for a billiard and card room on the north side. A large dancing-hall, with alcoves leading to the dormer windows, is over the other part of the house. It is intended that the rear porch form a feature of this structure. The piers are spanned by large arches, and it is so arranged that it can be enclosed during inclement weather.

The porch roof is flat, and covered with a wire trellis over which to train vines. This roof is on a level with the stair balcony. The entire end of this balcony is one great window, which throws light into the hall below.

In lieu of a porch in front, the front door is approached by a terrace.

The floor plans shown in Figs. 151 and 152 are arranged to bring the principal living-rooms on the south side of the house. The front reception-hall connects directly with the parlor and sitting-room, which are connected by sliding-doors, and can thus be thrown together so as to present a very agreeable prospect. The library on the north side is properly isolated. Back of it there is a side-hall and carriage-porch to the north.

The second floor is self-explanatory.

FIG. 153. — FRONT ELEVATION.

The little tourelles in the front form closets, and are lighted by small windows, as is shown on the elevation.

The front (Fig. 153), in the style of the late fifteenth century in France, is well adapted in its outlines to the artistic requirements of an American dwelling. All parts are of stone, yet there is sufficient of openness and fineness of detail to preserve to it the distinctively domestic qualities which must be inseparable from a structure of this character.

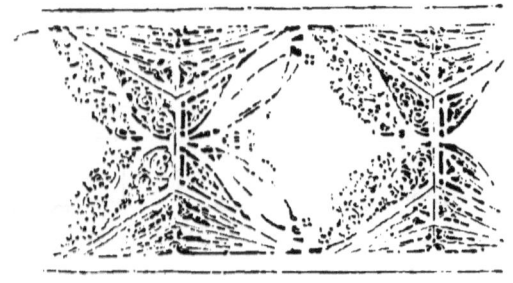

A BIT OF DECORATION.
(Louis H. Sullivan, Architect.)

SOME HOUSE PLANS.—*Continued.*

CHAPTER XIX.

BUILDING FOR INVESTMENT. — DEVELOPMENT OF AN INVESTMENT BUILDING. — IMPROVEMENT OF RENTAL STRUCTURES. — RAPID TRANSIT AND RENTAL PROPERTY. — ONE-ROOM APARTMENTS IN GLASGOW. — LASTING QUALITIES OF FRAME HOUSES. — THE REPAIR ACCOUNT. — THE DOUBLE HOUSE. — BRICK HOUSES. — A SMALL FRAME HOUSE. — RANGES. — THE BETTER CLASS OF TENANTS. — ROOMS OVER STORES. — FRENCH AND ENGLISH FLATS. — A SMALL KITCHEN.

BUILDING for investment might well be the subject of a volume. Most investment houses are erected in a short-sighted and careless manner, and in the end do not pay a fair return on the investment. There are a few notable exceptions to this statement. In some of the larger cities rental property is well constructed, but this is the result of expensive lessons from bad building. The first rental property erected in a new community is usually very shabby. The workmanship is poor; there is no regard for convenience and comfort. The occupants are not adequately protected from the heat of summer or the cold of winter. The sanitary arrangements are bad. In the course of time there is some improvement in general construction and convenience of arrangement. Still later it is discovered that it pays to put a bath-room in a house, but it is done in the cheapest possible way — a copper tub, a cheap water-closet, and the workmanship none too good. In the course of years even these houses are out-classed, and it is found necessary to have convenient, well-built, artistic structures. They may not be large, but everything must be substantial, well-arranged, and attractive. At first the cheapest, commonest rental-house in a new community pays well, but in a few years it becomes a bad investment. The well-arranged, though tawdry, house, with a few conveniences, makes it necessary for the owner of a less satisfactory building to rent it at a lower price. This in turn becomes true of the latter structures when the convenient, honestly built, artistic house is to be had.

The writer has seen large sums of money used in building cheap, showy houses, which, in a few years, become dilapidations. He has seen the savings and accumulations of wealthy citizens invested in a class of property which the heirs will find to be building-wrecks. The owners of this property imagined when it was built that they were realizing ten and twelve per cent. on the investment, while in fact a few years showed that their depreciation and repair amounted to five and six per cent. In building cheap property one must remember that one does not have to calculate interest upon one's money alone, but also upon a return large enough entirely to rebuild the property within relatively a few years. It takes an enormous gross return to pay a fair interest, fixed charges, and renew the principal. In truth, it is rarely accomplished.

The reduction in the cost of building, through the introduction of labor-saving devices, should be considered in connection with building for rental purposes. While the view of the situation from one decade to another will show that there is not a reduction in the rate of wages, there is a large relative reduction in the rate of profit to those who build. A two-dollar-and-a-half-a-day carpenter can do much more in the way of finishing a house to-day than he could two years ago. Improved machinery brings the woodwork to him in better shape for placing in position. It is more nearly finished when he gets it. The handicraft has largely gone out of the carpenter's trade. He merely erects what comes to him from the machine.

The cost of brickwork is being reduced, independent of the fact that the mason's wages have not been reduced. The brick is made by machinery, dried by steam, burned in improved kilns which utilize a very large proportion of the fuel, and thus the cost of the building is greatly reduced. The cost of iron and steel has been reduced by methods too well known to require mention. The same principle applies to all building materials. The cost of the labor which goes into all building products has been reduced by mechanical and systematic methods which we all understand in a general, if not in a particular, way. In truth, the tendency of all values in which labor plays any part is downward, for the reasons given. The relation of this fact to rental property is quite plain. One can build a better house for thirty-five hundred dollars to-day than one could ten years ago. The new house can be rented to the tenant who lives in the thirty-five-hundred-dollar house built in the last decade, for the same price which he has been paying, and give him a better house. This means that he who invested his thirty-five hundred dollars ten years ago must accept a lower rate of interest than the one who invests it to-day.

When there are a relatively large number of rental buildings being erected in a large town or city, there is always much talk about over-building. But one who observes must realize that it is never the new buildings that stand vacant. It is the older, inferior, out-classed structures which cease to bring a fair return on their original cost. This necessitates carefully considered, substantial, artistic buildings at all times.

The development of rapid transit naturally brings about great changes in investment property. There is less occasion for compact building. With rapidly moving electric cars, great distances can be covered within a few minutes. This means that very tall flats and apartment-houses need not be built in cities of moderate size, and that many that have been built will cease to be income-paying property. In the course of time, when the municipal authorities are wise enough to see that street-car fares are reduced, it will have a great effect even upon the tenement-house investment. It will not be many years before nearly every one can live in a house surrounded by a little plat of ground, and all because of the proper development of means of low-cost rapid transit.

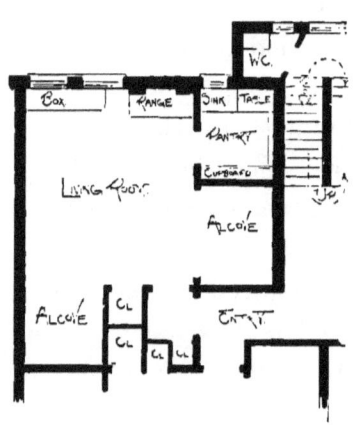

FIG. 154.— FLOOR PLAN.

The construction of rental property for those who wish only two or three rooms is the most difficult. Such people cannot afford to pay street-car fare at present prices, and must, therefore, be in a section of the city near their work. The city of Glasgow has solved this problem exceedingly well. The best exemplification of their plan is in the buildings in the region of the old Salt Market. There one finds even the one-room apartment conveniently and even attractively arranged. The buildings are large, substantial structures, four stories high. The first story is occupied by shops, and the others by apartments of one (Fig. 154), two, and three rooms, none being larger than three. The stairways leading to the various floors are of stone, the corridor walls are lined with enamelled brick, which is incapable of being fouled, and the floors of the corridor are of cement. The apartment floors are of well-laid oak; the wood trimming of the doors and windows is narrow in width and well painted. The range (Fig. 155), is set in tile. In front of it is a hearth and above it two shelves. As one enters the room, the impression is of an ample mantel and fireplace, rather than a range. Some of the cooking utensils are suspended against the tiling, and on the two upper shelves one sees plates, spice-boxes, and a clock. In appearance it is not greatly different

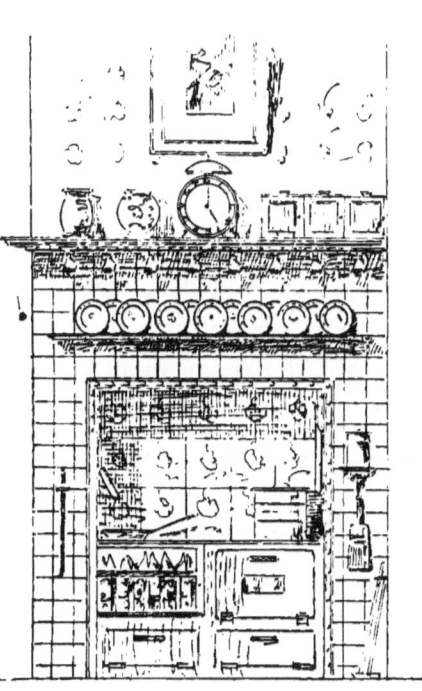

FIG. 155.—RANGE AND MANTEL.

from what one sees in a sitting-room or a dining-room of an American home of the better class. The plates and other articles of utility take the place of the bric-à-brac so common with us. Near this fireplace, as shown in the plan, there is a pantry. Under the window there is a sink, and under the sink a place for coal. Arranged around the walls of this pantry are all the conveniences of housekeeping that one could well imagine in such a place. Under the window in the main room is a window-seat and chest combined. At the end of the room farthest removed from the seat is an alcove for a bed. This, together with a closet and some shelves on the wall, completes the equipment of a very attractive apartment, which rents for three or four dollars a month. These structures were built by the city of Glasgow, to show the reluctant capitalists of that city what could be done in the way of providing proper accommodations for those who could afford to pay only a very modest rental, and at the same time yield a profitable return for the capital invested. The result from this standpoint has been entirely satisfactory. It is an interesting fact, in connection with the philanthropic work of that great municipality, that it is all adjusted on an interest-paying basis. One who notices these things must certainly have observed that philanthropic investments which do not pay a fair return upon the amount of capital and labor are certain of failure because of ultimate neglect.

Moderate-sized structures of either wood or brick, if they are but convenient and interesting, will usually pay a fair return upon the investment. The frame house is often considered temporary, for the reason that many frame houses have been built in a shabby, temporary way.

It has been my pleasure in this book to cite many examples of frame structures of a substantial character which have been occupied as income-paying property since the fifteenth and sixteenth centuries — three and four hundred years. A large part of Hildesheim and Halberstadt is of that period. These buildings are not only inhabited, but are in an excellent state of repair. One will find such old wooden structures scattered over Great Britain and the Continent. Some of the details which were given of the French timber buildings put to shame all of our modern ideas of carpentry. I do not expect our people to build frame structures as substantially as the old buildings of Europe. I cite examples of buildings of this character to indicate that very much better modern structures may be erected than those to which we are used. We may accomplish this relative result, and still not closely approach the excellence of the structures to which reference has been made.

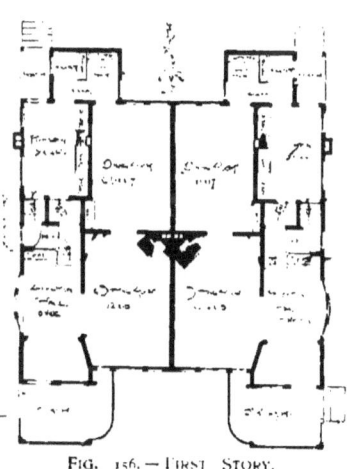

Fig. 156. — First Story.

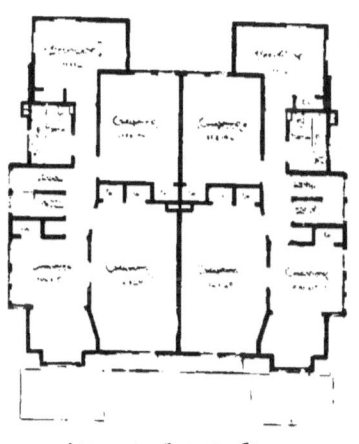

Fig. 157. — Second Story.

Our ordinary weather-boarded buildings are unsatisfactory after comparatively a few years. The joints bulge out, the boards get out of line, and the structures are altogether bad in character. If one were to use a solid seven-eighths-inch covering of the kind shown on page 153, one would find that it would last for a very long time without evidence of age. The framing of most of our wooden buildings is so carelessly done that we have no right to expect them to be permanent. The framework itself is covered with a common grade of sheathing, this with paper, and finally with the weather-boarding, or clapboarding, as it is called in many sections. Instead of the rough sheathing, with its open joints, if one were to use a good grade of solid flooring, with its tongued and grooved joints well driven down, running the boards on diagonally, there would result a covering worthy of very good framing. Afterwards this could be stripped and lathed with a metallic covering preparatory to plastering, or it could be cross-lathed and plastered; it could be covered with solid seven-eighths-inch siding, of the character previously referred to, or it could be shingled. All of these are permanent, and, properly treated, can be made very interesting. But this alone would not make permanency. The decorative features of these buildings should be developed so as to give no opportunity for water to get behind them and cause decay and eventual destruction. It is

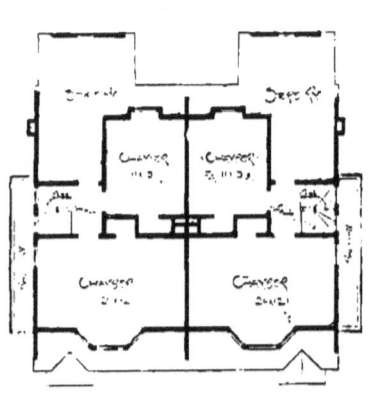

FIG. 158.—THIRD STORY.

entirely possible so to do all this work that it will stand the ravages of years without appreciable expense for repairs. The repair account on a great deal of investment property,

FIG. 159. — EXTERIOR.

particularly of the West, has been very large; and unnecessarily so, because of the unexpressed thought in the minds of many people, that great care and attention were not necessary for a rental house. It has ordinarily been true that nearly anything would rent; but as the country grows older this idea is dissipated, though the habit of building carelessly does not disappear at the same time. The very essence of a good investment of this character is to build something which does not require future expense. It is simply a question of good building.

The house shown by Figs. 156 and 157 has been built substantially as here shown many times, though the details in plan, as well as in construction, are here better developed than ever before. The stair-landing usually has been kept inside of the main lines of the hall. This has always made the stairway more or less

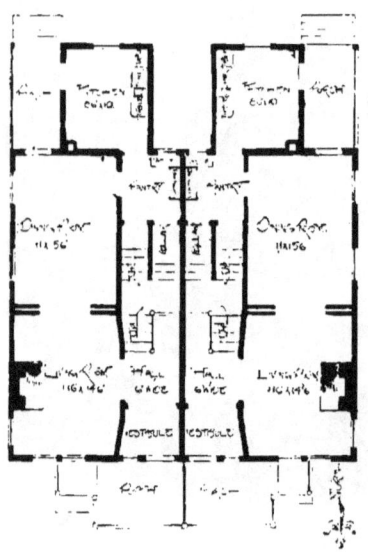

FIG. 160. — FIRST STORY.

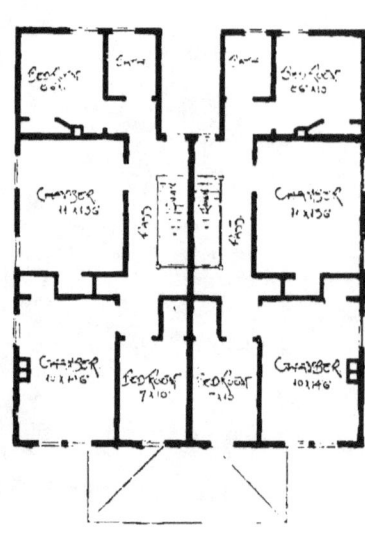

FIG 161. — SECOND STORY.

cramped. It has made it difficult to get furniture up and down, and while it is really better, where the mover is properly provided, to take furniture in through the second-story windows, there is a prejudice against this method. In this instance the stair-landing is all outside the main lines of the house; that is, it is overhanging. In other respects this plan is so much like many other plans which

I have made that I will not explain it in detail. There are two good rooms on the third floor. (Fig. 158.) There are none better in the building than these. The division wall between the two sides is of brick, and is eight inches in thickness. Many frame houses are built with a filling of brick, four inches in thickness, between the studs. This does not prevent the passage of sound. Nothing less than an eight-inch wall is satisfactory for that purpose. Fig. 159 makes plain the external features of this building.

The plan shown in Figs. 160 and 161 has six rooms and a bath on each side. The stairway is a combination affair, and, in a building of this kind, is satisfactory. The laundry apparatus in both of these houses is in the basement. The cellar floors are of cement, and as the rooms are well lighted from above, they are eminently satisfactory for this purpose. Two water-closets are provided, one in the bath-room and one in the basement. A privy building should never be thought of in connection with any city structure. Even if there must be only the vault on the outside rather than a sewer, it is best to have the water-closet on the inside of the building. If the plumbing work is properly done

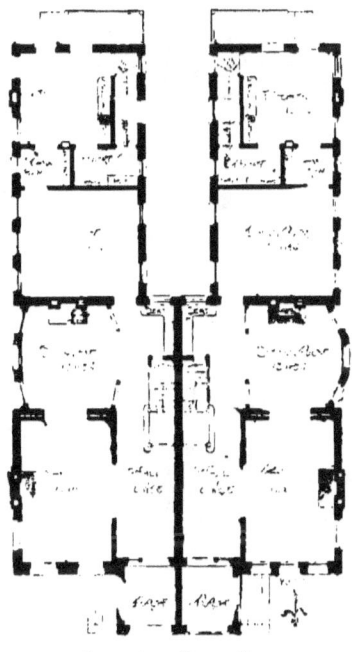

Fig. 162 — First Story.

according to a skilfully drawn specification, there is no more risk from a sanitary standpoint with a vault than there is with a sewer. Under any circumstances the danger

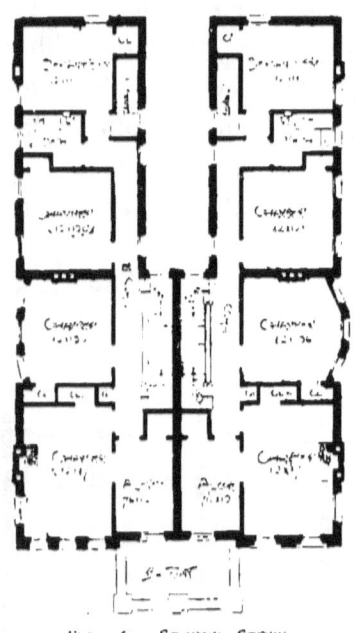

Fig. 163.—Second Story.

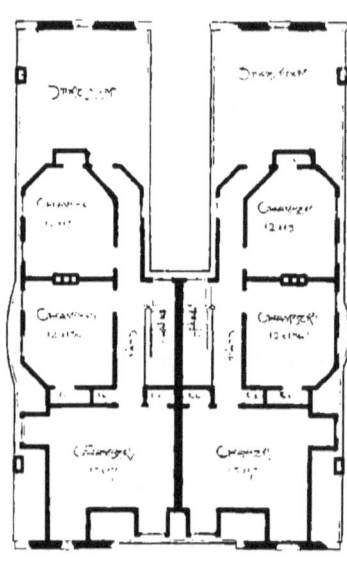

Fig. 164.—Third Story.

grows out of unskilful plumbing, and never from the vault or sewer alone.

Buildings of this kind should be finished in hard wood throughout. While it is well to paint that in the kitchen and pantries, all other parts of the house should have a varnish or hard-oil finish. The floors should be at least as good as narrow-worked (say three inch) yellow pine, and should have a proper floor-finish before the house is occupied. Thus it is not necessary for the tenant to purchase a large outfit of carpets or other floor covering than rugs, which may be used as well in one building as in another. If one wished to put down a carpet it could be done. It is not usual in the West to supply ranges. This should be done, whether it is usual or not. It is a measure of economy. The moving in and out of a heavy range is calculated to work great injury, and always causes direct expense in connection with the hot-water apparatus. A tenant will pay a larger rental for a house furnished with hard-wood floors and a range than for one

FIG. 165. — TWO DOUBLE HOUSES.

less satisfactorily equipped in this respect. In other respects the character of good building is always self-evident in general ways. Porcelain-lined tubs always command respect,

FIG. 106. — PARLOR.

and good general construction expresses itself in many ways. Add to this an artistic conception, and the building will never be out of style. Good taste and good sense are always in style.

In every moderate-sized city there is a demand for rental property containing a larger number of rooms than any of the houses which have been described in this chapter. The successful merchant who can profitably employ all of his capital in his business, the railroad superintendent or other officer who may be transferred to another city, finds it more desirable to rent than to own property. The number of such tenants is not relatively large. They seek the best that is offered, and are disposed to pay a fair price for it. Figs. 162, 163, and 164 are the floor plans of a double brick house of this character. The principal object of interest in connection with this building is the lighting of the main hall. It

FIG. 167.— SITTING-ROOM AND STAIR-HALL.

has windows in the rear, which are connected with an open court. Thus, in coming in from the front door, one faces the light rather than the darkness which is so common in structures of this kind. The dining-room is lighted from each side; that is, one side is on the court and the other on the outside of the house. This court, it should be understood, is open at the end, and of course at the top. It presents means of thoroughly lighting and ventilating all of the halls and rooms. The sitting-room in this plan is merely separated from the hall by a screen of open woodwork.

This building, as well as all others described in this chapter, is heated from a furnace in the basement. There is a laundry under the kitchen, and a bath-room on the second floor, which is well fitted with the best modern plumbing fixtures. The entire structure is provided with speaking-tubes, electric and gas light, — all of which makes a complete and satisfactory house. Before these houses were occupied, or even rented, all of the walls were properly decorated, the floors finished, and altogether quite as much thought and attention were given to them as could be given to a house which the owner intended to occupy.

The view of the exterior (Fig. 165) shows that there were two of these double houses erected — varying as to the exterior, but of the same floor plan. The one on the right hand was designed in the style of the early French Renaissance; the one on the left in that of the later Dutch Renaissance. All of the details were very carefully studied and executed, with a view of presenting two structures dissimilar in style and external appearance, at the same time that there would be no broad difference of opinion as to the relative merits of the two designs.

No one likes to live in a row of houses where all are alike. In these structures the style of the interior doors and of the woodwork, the hardware, the mantels, the decorations and all were varied, so that the impression of sameness and the idea of the duplication of parts would be dissipated as far as possible. It has been a source of great pleasure to the designer to be asked which one of the buildings was his work.

A great many of the rooms over the stores in small cities, and the outlying districts in larger ones, do not bring an adequate return on the investment, for the reason that they are not properly arranged for rental purposes. There is simply a division of space, indefinitely spoken of as rooms. A three-story building may be profitably rented as apartments if the exterior is made sufficiently interesting to command respect, and if the interior is well planned.

FIG. 168.—A STORE BUILDING.

Outside of the large cities, living in blocks and above stores carries a certain amount of stigma with it. This is largely true because such places are not fit to live in. They are not arranged for living, and their occupancy is associated with discomfort, if not with uncleanliness. But it is astonishing how quickly people change their minds when sufficient reason is afforded therefor. If property which so often does not pay can be arranged on a profitable basis, it will add just that much to the wealth of the individual and the community. In France and most of the large cities of Great Britain, nearly every one lives in apartments. This was primarily brought about by necessity, and the necessity has developed good plans. The fact of there being a store on the first floor does not detract from the high character of the apartments, providing their arrangement, convenience, and equipment justify such a rating.

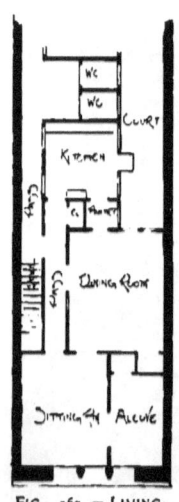

FIG. 169.—LIVING-ROOMS OVER STORE.

A French apartment cannot be used in America just as found there where they are nearly always over shops. The American store is a different thing from a foreign shop, the average depth of which is twenty-five or thirty feet. Back of the shops are the stairways, courts, and rooms for the concierge or care-taker of the establishment. The American store must be from sixty to a hundred or more feet in depth. Hence, apartments above stores must be adapted to new conditions. The elevation shown in Fig. 168 is of a building of this character. The entrance to the stairway and its relation to the other parts of the building are clearly shown. The second floor is approached from a stairway to the left of the store, and each apartment has its distinct doorway opening into a hall, which belongs to it alone. If the building were three stories, the stairway to the third story would be in the middle, and give the same arrangement on the third floor for both apartments that is shown on the second floor, excepting that the dining-room would be longer and the fuel-room better arranged than for the front apartment in the second story.

The small kitchen would disturb most Americans, unless they have lived in a French apartment and been comforted by a French kitchen, which is usually not more than five or six feet wide and eight to ten feet long. What is wanted in a kitchen is convenience rather than size. The range should be built in a large alcove, and a hood provided over it, with ample opening to the exterior, so that absolutely all of the heat can be drawn out. The efficiency of such an arrangement cannot be questioned. Joining the range are a table and a sink. The sink has hot and cold water connections. Above the head-height and around the kitchen are shelves, and projecting into the court is a little screened cooler, with a door in it.

In this plan (Fig. 169) we have a court beginning with the ceiling-joist of the storeroom, which should be entirely open at the top and covered with a tin roof at the bottom. If need be, a skylight could be provided to throw light into the store.

With proper arrangements for getting rid of garbage and for marketing, no one need ever leave the apartment excepting for business and recreation ; and the difference in housekeeping, as compared with that of larger establishments, is hardly to be estimated, excepting through experience. This floor plan and arrangement are typical of a very modest development of apartment rooms over stores. The situation chosen was with no direct light on either side, because of its difficulties. However, direct light is let into each room.

DECORATIVE MOTIVE.
(Louis H. Sullivan, Architect.)

SOME HOUSE PLANS.—*Concluded.*

CHAPTER XX.

PERSONAL CHARACTER EXPRESSED IN HOUSE-BUILDING. — A HOUSE FOR THE BRIGHT, CHEERY LITTLE WOMAN. — FOR THE EXACT, DELIBERATE BUSINESS MAN. — THE PICTURESQUE CHARACTER. — LEGITIMATE VARIATION OF DESIGN INFLUENCED BY PERSONALITY. — A MISCELLANEOUS COLLECTION OF HOUSE PLANS.

FROM the architect's standpoint, the designing of a house is a great character-study. He need not feel that he must pamper the taste of people. He may, however,

FIG. 170.

study their general disposition, satisfy their general requirements, and not depart from an artistic impulse. It would be quite absurd to expect all to build houses of the same character. The dignified professor does not have

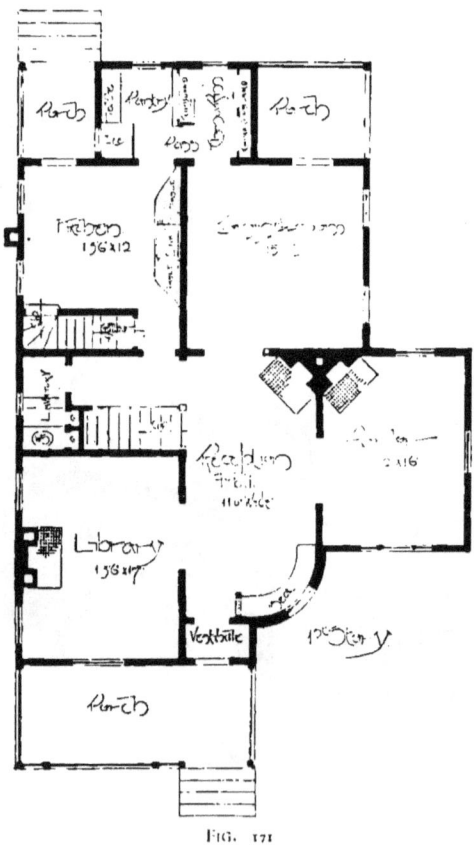

FIG. 171

the same taste in house-building as his less serious neighbor, whose disposition is moulded by a less serious calling. It would be altogether inconsistent and uncalled-for if the architect should insist upon clothing all his houses with the same character of design. A bright, enthusiastic, cheery little woman goes into an architect's office to talk about a house. It would be quite absurd to design for her a building similar to that which would be satisfying to a quiet, exact, deliberate business man. The architect need not be a puppet to meet all these conditions. He should have knowledge of architectural styles, as

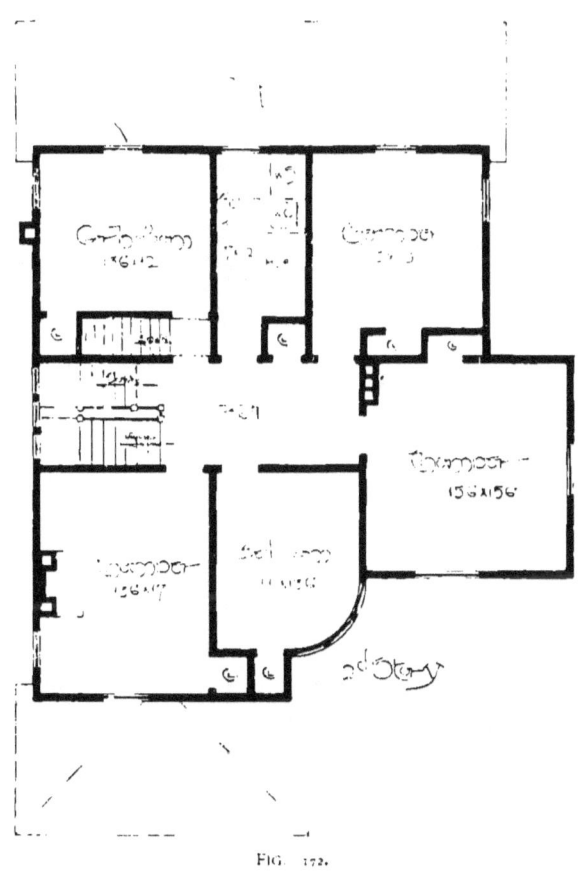

FIG. 172.

well as knowledge of character. The college professor may wish his house to be as unobtrusive as possible; he may wish it to attract little attention, to sit quietly in its yard. He may wish people to pass by it without thinking of it, without recognizing that it is a new house. It

FIG. 173.

is to be reposeful. The owner leads a quiet, studious life.

The bright and enthusiastic little woman who dominates her household is not pronounced in her taste, nor is she in search of public applause. She merely wants a home which is suited to her own happy moods, one which will interest her friends. Its details are all in good taste, it is quietly picturesque in character, has a hospitable porch, quaint, interesting detail, nooks and cosey corners in the interior, and all that goes to express bright, interesting social life.

There is another character, of an extreme but not unusual type, — the smart trader who has accumulated wealth rapidly, who has surprised himself, and who wishes the world to know the general result. He would express himself in architecture. What is the conscientious architect to do with him? This character steps briskly into an architect's office on a bright day in early spring. He wears tan shoes and a silk hat, not well placed; a purple necktie and bright kid gloves. An Old Colonial house will not do

FIG. 174

for him. His building must not sink back among the trees. He owns the best corner on the best street. It must be a good-looking house, and it must not cost too much money. Must the architect try to reform him? Such an effort would be without avail. Must he make a house to fit the tan shoes, the purple necktie, the silk hat, and the lake gloves? Not that. Here is a man of rather brilliant taste. He wants something picturesque. Are all picturesque things in bad taste? are they unworthy? Is it beneath the dignity of an artist to do the picturesque thing? Look to Brittany, to Normandy, Holland, and Germany, Switzerland and England, during the period of the Renaissance; look to the mediæval architecture. Give to this client something which is picturesque; in truth, rather brilliant in outline, and, when it comes to making the details, refine them to the last degree. The architect has satisfied his client,

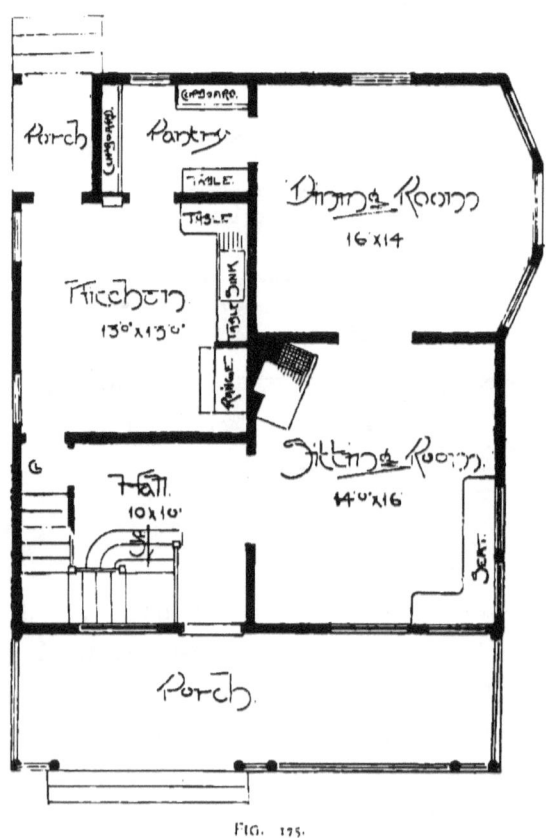

FIG. 175.

and in the refinement of detail he fully justifies himself. The wild architecture of most of the Western cities indicates clearly enough that the disposition of most people who build leads them to the picturesque. Most of them have only what is crude in outline, as well as what is obnoxious in detail.

The illustrations given are all of an extreme character.

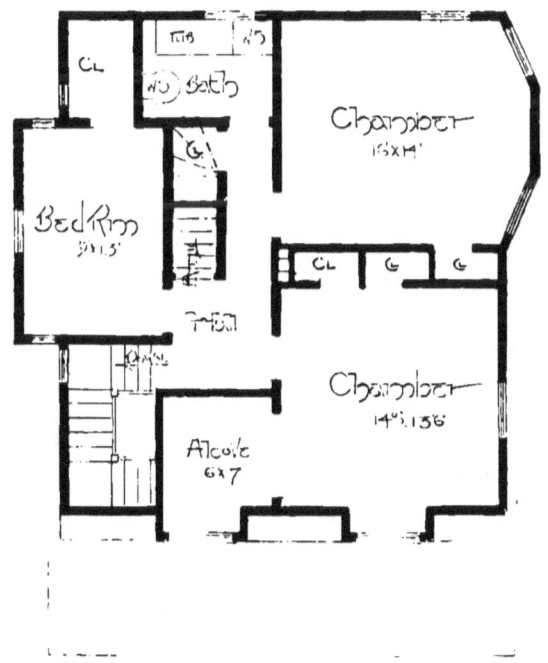

FIG. 170.

Every architect will recognize them. If he is wise, he is not annoyed or disturbed by any of them. Each is a type, and, as soon as he meets it, he knows exactly what to do. It is the effort to reform every one with whom one comes in contact which makes trouble. People do not go into an architect's office to be reformed. The architect can justify himself and his art along the lines which have been here laid out, and in so doing he need not feel that he is in the least debasing his art. There is no reason why he should make an inartistic outline or detail. If there is difficulty in satisfying a client either as to floor plan or general design, the architect may well take the blame to himself. It is altogether a question of resource on the part of the designer. No general problem is ever presented to an architect which may not be worked into an artistic success. It is a question of artistic resource and the study of character.

The plans which are pictured in this chapter were developed in a varied practice, and for them no apologies are offered. They are quite satisfactory to the people for

FIG. 177. — FIRST STORY.   FIG. 178. — SECOND STORY.

whom they were made. As floor plans they are all convenient from a housekeeper's standpoint, and when viewed æsthetically they are not ugly. The house represented by

Figs. 170, 171, and 172 is one of moderate cost and relatively large accommodation. There are four good rooms on the main floor, five bedrooms on the second

FIG. 179. — FRONT ELEVATION.

floor, three bedrooms and storeroom on the third. It serves the purpose of housing a large family at a relatively small cost.

Figs. 173, 174, 175, and 176 represent a compact house of low cost. The three principal rooms on the main floor are arranged with reference to the greatest possible economy of labor in housekeeping; there are three bedrooms on the second floor, and two good rooms on the third. Classic or Old Colonial details are used in the clothing and decoration of this structure.

Figs. 177, 178, and 179 are of a house of the general character frequently erected in country towns by the successful merchant, banker, lawyer, or other professional man. Bearing in mind that the general arrangement of the floor plan has been influenced by all of the principles of convenience which have controlled the planning of other structures described in detail in this

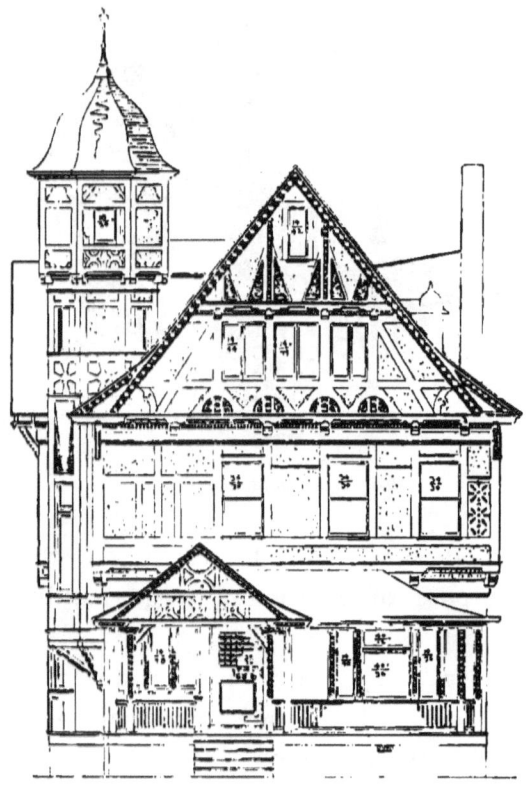

FIG. 180.

book, and in the one on Convenient Houses, it is unnecessary to go into a detailed description at this time.

Figs. 180 and 181 are of a house with South German outline and general form, though the details — that is, the mouldings and strictly decorative forms — are worked out from the Byzantine. Many of the details of the South German architecture of the sixteenth and seventeenth centuries are rather common in character; but by using the South German forms one can get a very beautiful outline, and with the Byzantine details can secure great refinement.

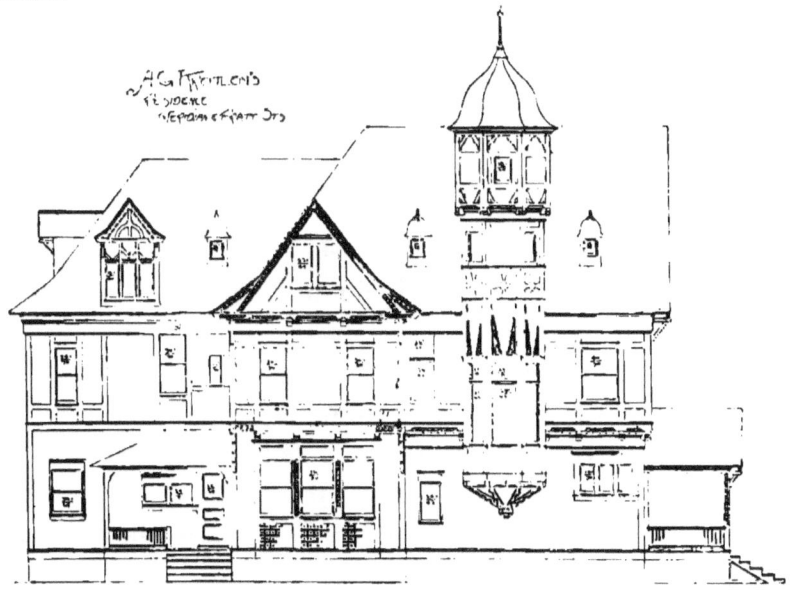

FIG. 181.

MATERIALS AND DETAILS.

CHAPTER XXI.

SHINGLE-HOUSES. — THE PROPER SURROUNDINGS. — THE STAINS OF TIME. — ARTIFICIAL STAIN. — EXAMPLES. — SLATE WALLS.

THOSE who like shingle-houses like them very much. Those who dislike them go to the other extreme. There are no moderate expressions from the public. A great deal of bad work has been done in shingles. This material should be considered merely as a medium for artistic expression. An artist may use charcoal, water-colors, oil, or pastel ; the artist house-builder may use shingles, brick, weather-boarding, or plaster. Each is a medium. The mistake in the use of shingles has been in placing them where they do not belong, using them in ways not justified by the character of the material. A shingle-house on a narrow lot and a busy street is out of place. A shingle-house should be associated with a green lawn, a little clump of bushes, old-fashioned roses, hollyhocks, and other picturesque flowers. A shingle-house crowded close to the sidewalk is misplaced.

People do not discriminate closely. They say, " Here is a new shingle-house ; it does not look well ; why were shingles ever used in this way ? " The improper use of shingles, in combination with typical planing-mill brackets, common turned columns, ugly jig-sawing, and over-wrought gables, always makes an ugly picture. In many communities such may be the only example of shingle-work.

FIG. 182. — OLD ROSES AND HOLLYHOCKS.

People of natural good-taste are repelled by it. Those who accept all fads adopt it.

There can be no question but that some of our best work has been done in shingles. Shingles left to their natural color or properly stained are infinitely better in color than paint. There is a natural lack of uniformity, and a richness of tone, in shingle colors which are not found in other materials used in wall covering. The nearest thing to it is the color found on buildings which have never been painted, or from which the paint is entirely worn off. The rich grays which come in this way are unexcelled by any of the artificial preparations. The use of stains, like any good thing, has been abused. Colors which should never have been applied have been used,

FIG. 183 — OLD CONNECTICUT.

and the search for novelties has done what it always does — crushed out the artistic. Some of the stains have been made to counterfeit the appearance of old age, but unsuccessfully.

There can be no doubt but that some of the best houses which have been built with shingles were unstained, and left to take the beautiful gray color which nature gives them. But most of us are too impatient to wait for nature's coloring. We have at our hand bleaching materials which are really of some service in hastening the natural process. Under any circumstances, the owner feels that in using this material he is really doing something which may be of avail, and is thus easily led into doing the right thing. This bleaching compound is as near nothing as one can well imagine. The shingles look about the same after it is put on as they did before;

FIG. 184. — ONE OF RICHARDSON'S HOUSES.

but it satisfies a house-builder's active impulse, and thus leads to more patient waiting. The writer does not propose to decry the use of stains, but wishes to show that some of the best work has been done without them. Some of the light-brown stains give colors which improve with time, and are very beautiful.

The old Connecticut structure (Fig. 183) is no

FIG. 185. — A CITY PICTURE.
(Louis H. Gibson, architect.)

doubt one of those examples which led to the use of shingles in house-building. The situation, the fences, the stone walls with their splendid colors, the bushes, trees, and the surrounding country, are exactly right for this kind of a house. This is one of those rare instances where the house forms an agreeable part of the landscape. These shingles were the old-fashioned hand-split shingles, and one can well understand, by examining this picture, that they have been on these walls for many years, and that their color is not comparable with any artificial product.

This picture of one of Richardson's houses (Fig.

184) indicates a fairly successful treatment of a city shingle-house, though, with brick and stone houses around it and electric-cars near it, one feels that it is misplaced. Not but that it is beautiful, but it is not in the right company. The shingles of this building are laid with a little more surface to the weather than is common, and on this account alone it is more pleasing than some other examples. The surfaces are somewhat broader and

FIG. 186.

Fig. 187. — A Seaside Picture.
(Arthur Little, architect.)

less disturbed by this means. The reader may profit by noting the simplicity of detail, or rather the almost complete lack of detail, on this building. It is good composition, without any attempt at decorative effects.

The little house on the corner (Fig. 185) looked better without its neighbors than with them. It was built several years ago, when the surrounding lots were vacant, when the street in front was not quite so prim as with its present asphalt covering. There were trees on the vacant lots, and altogether the surroundings were more congenial. It is interesting to notice the beautiful shadows which the square projecting roof casts on the circular corners. In summer, when the vines grow up near the veranda-posts, they help to make an interesting picture.

The training of an artist shows in this house (Fig. 187). The broad lawn, the low, flat building, with its simple detail, have just enough of the picturesque to make

FIG. 188 — IN CALIFORNIA.

a picture. The architect has worked out the composition exactly as a painter would produce a picture, and the lawn, the building and general surroundings, form one complete motive. The details are very simple, and very fine and delicate. It is not the kind of structure to attract the attention of the ecstatic, and this is one of its great merits. It is a natural house, and forms a part of one of nature's pictures. One almost wishes that the large projecting hood had been left off the doorway; yet, with all the other things which are so good, one cannot express dissatisfaction. This feature shows, as clearly as possible, the influence of bad example, even upon an educated architect and a trained artist. At the time when this house was built there was quite a reaching after these strained effects, and this half-dome-shaped hood over the window is the result of the impression made by these bad examples. It shows how difficult it is to resist such influence.

Slate is used, much as we use shingles, for covering houses in the smaller towns and cities of France. There is an example of such a structure on page 43, taken from Lisieux. In the old mediæval city of Vitre, in Brittany, there is a very large number of these slate-covered structures. At other times one finds the slate used to cover the timbers of the half-timber buildings. There are many examples of this kind of work in Rouen. Slate covers the braces and uprights, while the space between is filled with concrete. These things look quite well with their mediæval surroundings, and appear to fit the tradition which goes with them, but are not appropriate to the modern city. They may furnish us with inspiration, but not with examples for transportation.

It may be interesting to know that shingles make a warmer wall than weather-boarding or other wood-covering. Their laps are greater, and the manner in which they are placed gives a decided advantage in warmth.

In speaking especially of shingles in this way, nothing could be farther from the thought than to give this material undue artistic value. It has great value when properly situated, but is merely one of many mediums — brick, stone, terra-cotta, clapboards, cement. Beautiful houses are constructed in all these materials.

FIG. 189. — NEAR PHILADELPHIA.
(Wilson Eyre, architect.)

MATERIALS AND DETAILS. — Continued.

CHAPTER XXII.

FIREPLACES AND MANTELS. — HISTORY. — MANTELS OF THE RENAISSANCE. — DUTCH MANTELS. — MODERN MANTELS. — CHARACTER IN MANTELS. — TILE FACINGS. — ONYX AND BRICK.

NO doubt the first fireplaces were for cooking. Certain it is that those which first took the form closely approximating those now in use were used for that purpose. During the twelfth and thirteenth centuries in France there were many kitchens of the form indicated by Fig. 190. This was the great kitchen of the Abbey of Fontevrault. Drawings of many others of these early kitchens exist, but this one is a typical illustration. Many were more complicated in plan, but the general idea of all was the same. There was the building of circular form, with niches forming fireplaces around the wall, and the ventilating duct over the centre. During the fourteenth century quite a change was made in the kitchen. It became a part of the general structure, and the fireplaces were nearer the form shown in Fig. 191. As seen, it is constructed of stone, with details of an interesting character. The projecting mantel is high enough to permit one to walk under it. At the rear, and suspended over the fire, are iron hooks, from which cooking utensils can be suspended. This fireplace is from the Abbey of Blanche de Mortain.

The great rooms and halls of this time were heated by large fireplaces. Fig. 192 gives a double fireplace, from the great room of the Château of Courcy. It is splendidly sculptured, and no doubt formed a part of a superb picture. Great fireplaces of this kind were by no means rare. They are found in the Château of Pierrefonds and elsewhere. Some of them are spoken of as being long enough to receive logs seven or eight feet long. In the end of the great room at Poitiers there is a triple fireplace. It is magnificent in detail, and, with the great windows above it, forms one of the most interesting subjects of its kind.

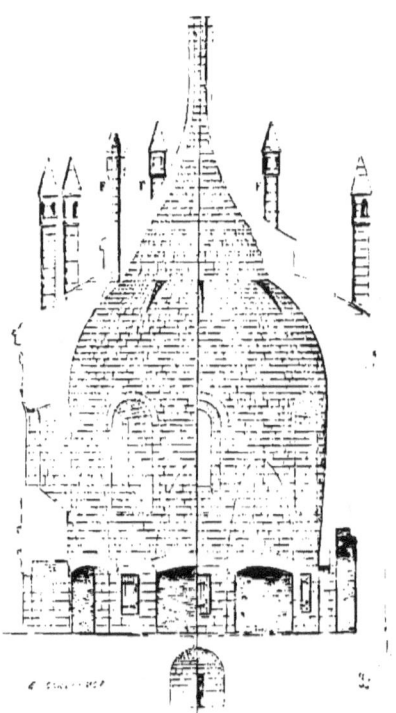

FIG. 190.— A THIRTEENTH-CENTURY KITCHEN.
(From Viollet-le-Duc's "Dict. de l'Architecture.")

The modern fireplaces of Brittany and Normandy are of the same type as that shown on page 58. To-day all of the cooking, unless it be the bread-baking, is done in these fireplaces. In this respect that part of the world has not moved since the fourteenth century.

Many of the most beautiful fireplaces of the fourteenth,

fifteenth, and sixteenth centuries were in the palaces and châteaux of France. There is room only to give views of one from Pierrefonds (Fig. 193), one from Blois (Fig. 194), and one from Cluny (Fig. 195). Two from Holland are also given (Figs. 196 and 197). That from Pierrefonds was of the early fifteenth century; those from Blois and Cluny of the sixteenth century. There can be nothing more interesting than these magnificent structures of the Gothic and early Renaissance periods. There are suggestions in all of them for our every-day use. The principal part of the construction of those already given is of stone. Many, however, have only the lower pilasters of stone, while the upper part is of wood and plaster. This indicates clearly enough that we may be able to do something in plaster in some of the forms indicated, and thus adapt them to our own immediate uses. While there would have to be some changes in modernizing the fireplace, these mantels are full of

FIG. 191. — AN EARLY FRENCH FIREPLACE.
(From Viollet-le-Duc's "Dict. de l'Architecture.")

suggestions, and can be readily adapted to our own time. On page 47 is given a mantel the outline of which was taken from a Dutch example. The details, however, are taken from French work, and a mantel altogether serviceable in arrangement and picturesque in outline is secured. Fig. 198 is a suggestion for a hall mantel. Some of the foreign examples we could not expect to duplicate if we so willed. We could hardly expect to find a sculptor so trained as to be able to execute those selected from the Francis I. part of the Château of Blois or the one from Auray. The latter is now in the Cluny Museum in Paris. They are great products of a great time, and while many are to be found in France belonging to this period, those given serve as typical examples.

FIG. 192.— CHÂTEAU OF COURCY
(From Viollet-le-Duc's " Dict. de l'Architecture.")

The Dutch mantels are selected from examples in the National Museum at Amsterdam. They were removed from old Dutch houses, and, together with their surroundings, were placed in this great museum. They belong to the seventeenth century. These particular examples are

Fig. 168.—Thiepval Chapel of Remembrance.

FIG. 105.—FIREPLACE IN CLUNY MUSEUM.
(Jean Goujon.)

interesting and valuable to us, because they are capable of being readily adapted to our own houses, and may be easily executed by our own mechanics. The lower parts

FIG. 196.—FIREPLACE IN MUSEUM AT AMSTERDAM.

are of tile; the decoration at the sides of wood; most the upper part of plaster, with wood for decoration. The mouldings belong to the Roman type, though they were interpreted by the Dutch artists of the sixteenth and seventeenth centuries. Naturally the decorative work as a whole partakes of the character of the time and place. These early Dutch mantels are particularly valuable to us, because they so readily lend themselves to present uses.

In the end of the dining-room described in chapter sixteen, the recess effect is secured in the manner shown by the plan. Practically it amounts to studding off one end of the room to the extent shown. The girder comes far enough down from the ceiling to admit of its being cased, and to provide space for china decoration immediately above it, after the manner indicated by the drawing. All of this room is wainscoted to the height of five feet with a perfectly plain surface of oak veneer. There are no panels, recesses, or decorations of any kind, excepting one moulding at the bottom and two at the top, one of the latter being about eight inches below the other. The entire wainscoting merely shows the grain of the wood. This wainscoting goes around the recess which is formed for the mantel. The lower member of the moulding referred to is on the level of the shelf, and the upper member is the top of the back of the mantel. Above this will be placed a picture flush on the chimney breast, painted by a capable artist especially for this place. The glazing of the windows on each side of the mantel is of simple forms of leaded glass, of the general character described in chapter twenty-eight. The tile work of the mantel is gray in color, an inch in width and four inches in length. All are unglazed, and the same form and color are used for the lining and the hearth. A narrow brass strip not greater in diameter than a lead pencil surrounds the grate opening.

FIG. 12.—FIREPLACE IN MUSEUM AT AMSTERDAM.

Other mantels are shown in this book. For the most part they are relatively simple in character, but not always because there was particular occasion for being careful in the use of money. Many of us are used to seeing mantels which are monumental and gross, the artistic value of which is largely estimated by their cost. They dominate everything in the room. If this were because of their beauty it would not be so bad, though even then it would not be justifiable. It is well to bear in mind that a mantel is an inherent part of the room which it is in. As few rooms in American homes are sumptuous and rich, there is no reason why the mantels should be monumental.

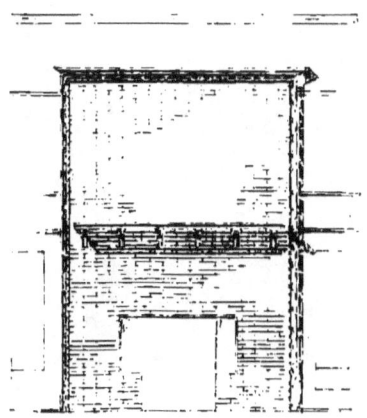

FIG. 198.—A HALL MANTEL.

The style of finish, the mouldings, the decorations and general character, should be the same as shown in the other parts of the room. There should be the definite relation between the mantel and the other part of the finish that there is between the different parts of a picture. It is not possible to have designs made for the finish of a house by the architect, and then buy a mantel which will be in character with the general design. The architect should design the mantel; otherwise the room is a hodgepodge — a bit of patchwork — a failure. The amount of this sort of thing which has been done is astonishing. Really intelligent people have plans made for a house, often costing a large sum of money, and then buy their mantels from stock designs. The result is before us in a large number of the houses which we see.

While the general character of the finish should in a measure influence the design of the mantel, there may be the same kind of finish in the different rooms, and yet a certain variety in the form of the mantels themselves. For example, in the parlor there may be a mantel with a mirror; in the library, one without a mirror; in the sitting-room, one with a broad expanse of tiling; and in the dining-room, an extra amount of shelving and devices for the display of china: yet the character of the mouldings, the general spirit of the design, may be in harmony with the details of the other finish, the casings, doors, windows, etc. While there is variety in general form and outline, it does not necessarily imply that there should be a particular kind of mantel for parlors, another for a library, or that there are mantels which are fitted only to a reception hall.

FIG. 199. — IN THE RECEPTION ROOM.

However, there may be such variety without keeping within prescribed lines.

The parlor is naturally more formal than the sitting-room or library. It is to be expected that a parlor mantel should be in character with such a room. It should be dignified, relatively simple in outline, though it may be rich in detail. One is generally disposed to indulge one's fancies somewhat more in a mantel for a library or sitting-room. Still, it is well to have it in mind that such a design should not be so strange or unusual in character as to be unrestful. It is something to be lived with, and we frequently see such designs tending towards severity or extreme simplicity. This is better than designs which are liable from their complexity to become wearisome when they are every-day companions. Naturally a bedroom mantel should be simple in outline and very refined in detail. One should be quite conservative in designing mantels for sleeping-rooms. The very tall mantel-shelf is a pretty form for a reception room. In this connection illustrations are given which indicate the design and the general proportion, though naturally they cannot be expected to show the coloring, or even the care with which the details of construction have been developed.

When the wood mantel came prominently into use a few years ago, there was usually a very large margin of wood and a very small quantity of tile around the frame. Excessive firing of the grate, and the unusual strain to which wood is naturally subjected when used in a mantel, led to an increase in the amount of tile used and a decrease in the amount of wood; so that it is now quite common to have only a narrow margin on the outside, the shelf above, with a small amount of wood surface above and below the shelf. In other instances we find mantels which are made entirely of tile, even to the shelf, and yet others of terra-cotta and brick. There is often, however, a heaviness in a brick mantel which is not altogether satisfactory. A very small number of the manufacturers of brick for such uses have sufficiently refined the details thereof to make them satisfactory for use in the rooms of an ordinary American dwelling. It is true, however, that many brick mantels are used. It is also true that many coarse and wrong things are done every day in house-building. While I do not absolutely condemn brick mantels, it is ordinarily true that they are too ponderous and heavy for the rooms in which they are placed. This is particularly true of red brick mantels. Bricks of other colors are now made which are more satisfactory for such uses. They are frequently very interesting in themselves, but when taken as a part of the room they lose their artistic value.

Enamelled tiles are the most satisfactory for mantel facings. One can get almost any color one chooses. The material is non-absorbent, practically indestructible, and in no way suffers from ill-use. Enameled tiles preserve their original color, form, and texture against all ordinary wear. They are relatively inexpensive, and there is no material of greater artistic value.

One is able to get these tiles with beautifully decorated surfaces, though unless selected by an artist of unquestioned ability it is best to choose simple coloring without any attempt at decorated surfaces. It is best to have not more than one color for the mantel-facing and hearth.

Glazed tiles are not nearly so interesting as the enamelled tile. There is a flatness in their coloring and a general commonness in their appearance which render them unsatisfactory.

Unglazed tiles are very beautiful, and but for the difficulty of keeping them clean they would be recommended without question for their artistic value. The darker unglazed tile may be successfully used, but the white, blue, and the light grays suffer greatly from use.

Mosaic designs especially require the guidance of the artist in the composition and use. The mere name mosaic, and the mere use of small pieces of tile as a novelty, is liable to lead many people to use them for mantel decoration.

Onyx facings and those of other special materials are obtainable. It may be said in general terms that many of these materials first came into prominence because of their cost. One is liable to be led in the direction of the unusual by something of this kind. Onyx, marble, iron, or other materials may be successfully used. More often than otherwise, however, these materials are selected independent of their relation to the mantel or the other parts of the room.

There should be a definite relation between the color of the mantel-facing and hearth and the color-scheme of the room itself. However, in ordinary house-building this is rarely considered. Mantel colorings are selected without regard to the decorative material or the decorative idea of the room itself. Before making a selection of mantel-facings of any material, there should first be an understanding as to the color-scheme of the room. It is necessary that this should be quite exact, otherwise a slight difference in shading would disturb the harmony of the arrangement. When we think that the mantel is nearly always selected without regard to the other woodwork of the room, and that the tiling or other material used for the facing is selected entirely without regard to the decorative plan, we can readily understand that nearly all rooms are failures from an artistic standpoint.

It is the practice of the writer to use very little mantel trimming around the fireplace opening. Often it is simply a very narrow metallic band. A fireplace-lining may be of tiling, cast-iron, or brick. The former is the most interesting and the most expensive. Cast-iron linings are now made from very artistic patterns. Fire brick take to themselves the beautiful coloring which comes from use. There has been great difficulty during former years in getting a tile lining which will stand great heat and hard usage. However, they are now made with a dovetailed back, which, when properly connected with the backing, makes a permanent and satisfactory lining.

The different forms of openings, grates, fire-logs, and irons are legion. All these things are now made in very artistic forms. Their successful use lies in the ability to make a proper selection.

MATERIALS AND DETAILS. — *Continued.*

CHAPTER XXIII.

DOORS. — THE DEFENSIVE. — HOSPITALITY. — MATERIAL. — FOREIGN EXAMPLES. — DOMESTIC DOORS.

ONCE upon a time doors were a very serious feature of all structures. This was true during the Middle Ages, when it was expected that a door would be a point of attack. Then it was not a question as to the hospitable appearance of the door. The sentiment was altogether different. The question was as to whether one would feel safe if one were behind the door. The owner of the great château felt that he was protecting

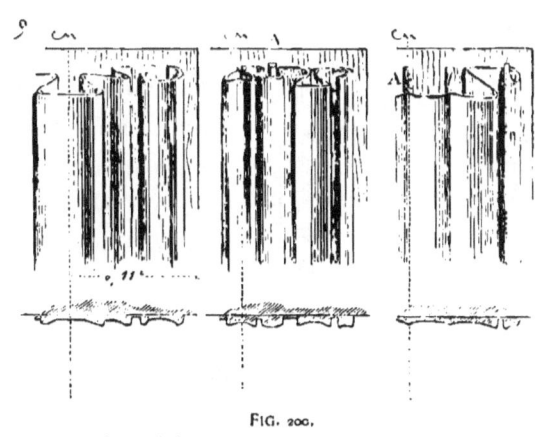

FIG. 200.
(From Viollet-le-Duc's "Dict. de l'Architecture.")

his guest and conveying to him a sense of security if he were able to pass through a door that was large and substantial as well as decorative. It was a peculiarity of the feudal times that in no structure, however serious its character, was the artistic neglected. We still find as remainders of that time doors massive and beautiful.

If the doors of to-day express anything, it is hospitality. The idea of great strength has no part in the sentiment of a modern door. If one were to consider the entrance door to a modern house ideally, the thought would suggest itself that it should reflect the general character of the occupants. In fact, however, this is not often the case. We find doors loaded with very common detail and fitted with glass often too brilliant and inharmonious in color. We may find very clever people, with good manners and great dignity, behind very crude and pretentious doors. Nevertheless, when one of good taste and natural knowledge builds a house, we do not find anything of this kind in crossing his threshold. There is an expression of quiet taste, good manners, and hospitality in the door. We recognize at once that it is substantial, that it will keep its place for many years, and that in its mouldings and other decorative details there are the marks of the hand of an artist. We naturally enter such a house prepossessed favorably.

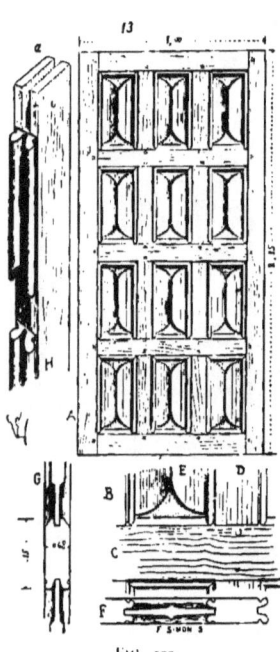

FIG. 201.
(From Viollet-le-Duc's "Dict. de l'Architecture.")

Many doors have their wood surfaces entirely cut up and covered with what is supposed to be ornament. It costs money, and the owners are told that it is a new thing in the way of doors, as much as to say that it is the style this year. For instance, the builder asks an architect: "What kind of wood do you think will be used for doors and finish next year? I am going to buy some lumber, and I do not wish to be out of the style." Without expressing all that is in his mind, the architect tells him that he thinks quartered oak will be good material to buy. "I am afraid not," is the reply; "I have used a good deal of oak during the last two years, and I am quite certain that it will not be used next year." When one bears in mind the many hundreds of years that this material has been in use and the great work which the artists of all centuries have done in it, one realizes that there can be no such thing as fashion in a material of this kind. One often sees painted poplar and pine entrance-doors quite covered with expensive millwork, all of which is as ugly as it is possible for ignorant labor to make it. How much better it would be to use absolutely plain hardwood doors, nicely finished, rather than these

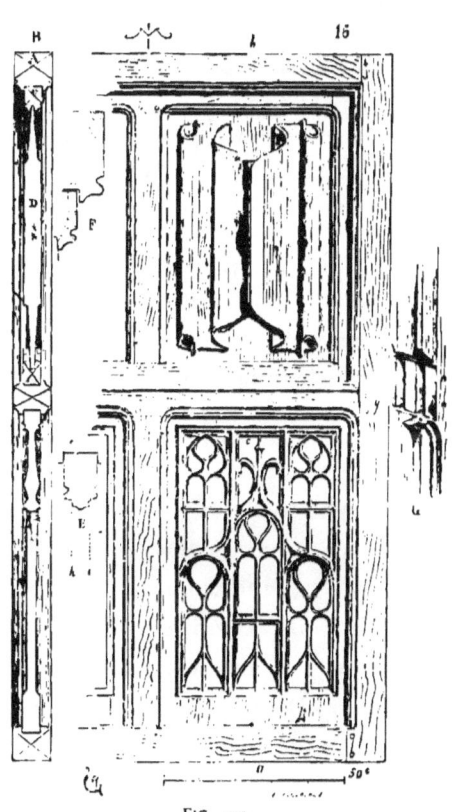

FIG. 202.
(From Viollet-le-Duc's "Dict. de l'Architecture.")

expensive, inartistic productions of the more pretentious kind.

The general principles which apply to the designing of entrance doors can be applied to those of the interior. Primarily, it is well to remember that the grain of the wood is apt to be more beautiful than anything we can add to it, and the forms which we may add, if they be not skilfully considered, detract from the interest of the natural surfaces.

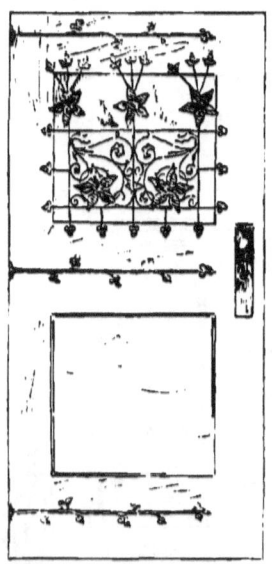

FIG. 203. — A FRONT DOOR.

Figs. 200 and 201 indicate what may be done to add to the beauty of the natural surfaces in handling the wood. There is nothing particularly interesting in the panel forms themselves, but the way in which they are cut adds to the variety and interest of the natural surface of the wood. This is true also of Fig. 202, which is somewhat more elaborate in detail than the others, but probably has no more artistic value than they, though the different forms given to the exposed surfaces add to their interest.

Fig. 203 indicates a doorway which depends almost altogether on the breadth and richness of the wood surfaces for the interest which it presents. It requires particular skill and special knowledge to construct a door of this character. The panels are flush, and the variety of surface is secured entirely from the character of the veneers which are attached to the structural parts. The iron grille which is fastened on the outside of this doorway is wholly decorative in character. We cannot imagine that the protective influence of such a piece of construction would be either necessary or effective. It has, however, quite a decorative value in this instance.

The door in Fig. 204 has a single large panel, with a small glass panel placed in the upper section. The rails and wood-surfaces of this door are quite broad, though the details are more complicated than those of Fig. 203. The former is more satisfactory. However, the latter is in character with the other finish of the house, and satisfactory in its relation to the other parts.

FIG. 204.

Fig. 205 is another of the plain-surfaced doors. The hinge-plates on Fig. 206 are solely decorative in character, having absolutely no constructive value. They add nothing to the strength of the door, though it must be confessed that they add decorative interest.

The door in Fig. 207 comes from the very interesting city of Saint Malo, in Western France. It was executed in the seventeenth century. It is quite complicated in design, yet, being the work of a true artist, it is a great success. He, no doubt, studied this door in the same serious, painstaking spirit, and with the same measure of enthusiasm, which is

FIG. 205.

common to one of our great painters to-day. It is such a spirit in our architects which will develop a better architecture for us.

In presenting designs for doors in this way, one cannot feel that he is doing his full duty unless he impresses upon the reader the idea that he cannot select from such designs in the same spirit that a lady would make a choice from a pattern book. These doors have their proper place only when properly used — with their right surroundings. They are selected from actual examples, and are referred to here as a means of demonstrating what has been done.

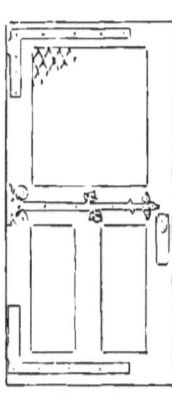

FIG. 206.

FIG. 207. — A BRITTANY DOOR.
(From "La Vieille, France," A. Robida.)

The door shown in Fig. 208 is decorated with nail-heads. These nail-heads are now made in iron and bronze, after good foreign models. It is proper to say in this connection, as has been done before, that the nails are not necessary to the construction of the door, and from certain artistic standards it is questionable whether they could be used in this way. However, if we were to eliminate all that was not structural and depend entirely upon our decorative construction, we should at least find ourselves very severe critics.

The inside doors given without special reference are those which may be used in a house of the highest or the lowest cost. They could be of pine, poplar, quartered oak, bird's-eye maple, birch, mahogany, or other kinds of wood. A great deal is lost in interior finish by the use of elaborate detail. Not many years ago it was often thought necessary to add to the complexity of a door-casing in proportion to the amount of money which an owner wished to invest in the building. If he were a very wealthy man, and were building a very pretentious house, the door-casings expressed his ambition. They were ten or twelve inches in width, and composed of many mouldings. While the tendency is away from this sort of thing, we have not fully realized that relatively plain casings have a much greater decorative and artistic value than those which are so covered with mouldings as to obliterate all plain surfaces and the natural beauty of the material used. It is well to bear in mind that the work of the architect and decorator should go together. In truth, under ideal conditions, the architect should control all decorative work. This being true,

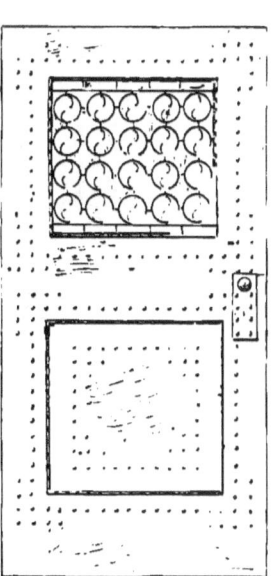

FIG. 208. — DECORATED WITH NAIL-HEADS.

the woodwork should not overshadow all other interests. There may be a variety without grossness, and yet in harmony with the decorative spirit of the interior. The doors, windows, the woodwork which surrounds them, and the mantel are merely parts of a decorative scheme.

FIG. 209.

They should be in harmony with it. The decorative idea and the character of the woodwork should go hand in hand. The examples which are given convey the intended impression quite as clearly without additional matter as they would with it. It may only be necessary to repeat again, for the sake of emphasis, that one cannot open a book and select patterns and designs for finish or other detail and expect to produce a successful interior. There

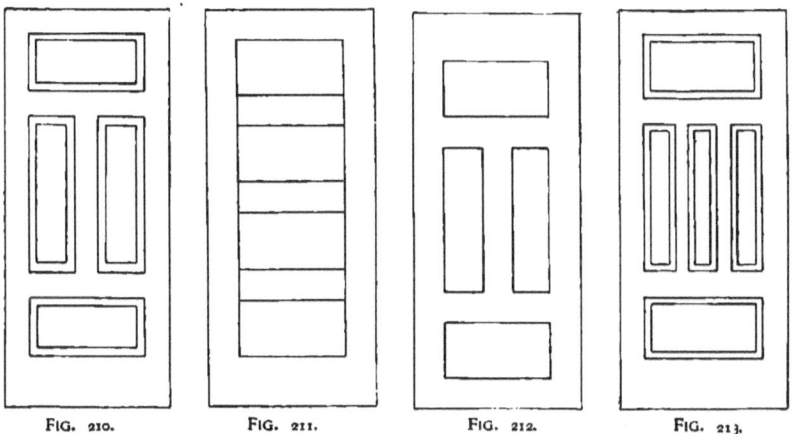

FIG. 210.   FIG. 211.   FIG. 212.   FIG. 213.

must first be a general idea as to what is intended, and then the details must fit the plan.

FIG. 214.

MATERIALS AND DETAILS. — *Continued.*

CHAPTER XXIV.

STAIRS. — FOREIGN EXAMPLES. — BRITTANY. — FRANCE. — HOLLAND. — GERMANY. — BROAD LANDINGS. — OLD COLONIAL STAIRWAYS. — IRON RAILINGS.

THE decorative motives in architecture are derived from the useful. There are few practical features of a house which can be rendered more interesting than a stairway. Yet if one without previous knowledge were selecting a motive for decoration, one would hardly take the form of a stairway. We take a form developed by necessity, decorate it, and if the work be properly done we say that it is artistic; all depends upon the decorative handling. I have seen beautifully carved ladders in Brittany, where they frequently do service as a stairway. In that country we find a people poor in this world's goods, yet rich in artistic surroundings. There are only two things needful to artistic production — artistic appreciation and the artists. We, with all our wealth, rarely reach the artistic; on the other hand, we see others attaining it without the wealth. We need artistic guidance.

When we see what the rest of the world has done, we find ourselves less timid in making departures from the beaten path. I have made a number of selections of stairways from foreign examples. It is not expected that with any other number of examples before us we should immediately go and do likewise. It is by the contemplation of what others have done that we are led to make departures from a well-worn path. The trouble with many of the stairways in our houses is not in their plan of arrangement, for in this many are picturesque and quite pretentious. It is merely a question of the refinement of the detail, the careful study of the parts intended to be decorative. The plans of stairways in general are picturesque enough, and there is enough labor and material used in their construction to make them beautiful, if the labor were only properly guided into artistic lines.

In many of the structures of the Old World the stairways are built in towers which extend into the main court around which the building is constructed. This is true of many of the military and civil structures previous to the fourteenth and fifteenth centuries. While many monumental stairways were built during this period, the very large majority were simpler in character. They were usually of stone and wound around the central pillar. There are several stairs of this kind in the house of Jacques Cœur at Bourges.

FIG. 215.—A BRITTANY STAIRWAY.
(From "La Vieille France," A. Robida.)

One enters the courtyard and then passes into one of the several stairways depending upon the section which one wishes to enter. Instead of communicating with the different parts of the structure from halls and corridors, they are connected with the court on the ground, by as many stairways as are necessary to afford independent connection with the various apartments. The stairway towers in the court of the house of Jacques Cœur are quite decorative, being picturesque in outline and rich in detail. It is only in the sixteenth century that we find stairways broad, liberal, and sumptuous. The Château of Chambord has a magnificent double stairway surrounded by grand corridors which lead to the magnificent tower which is the great feature of this wonderful structure. This stairway is so arranged that two people starting upon opposite sides may have each other always in view, though they cannot come together until they reach the main floor. They wind around over or under one another.

FIG. 216. — THE LANDING.
(Sketch by A. Robida.)

But the examples which we select are of a more modest character. They do not belong to great establishments like this house of Jacques Cœur at Bourges, or the great château built by Francis I. at Chambord.

Fig. 215 is a stairway in the corner of a hall in one of the quaint structures of Morlaix in Brittany. This stairway is placed back of the front tier of rooms, and winds to the top in the way shown. Galleries lead to the various rooms of the structure. The great hall is lighted from above by skylights. This stairway belongs to the sixteenth century. The construction is in oak, and many parts of it are literally covered with decorative sculpture. That around the cabinet on the main floor is particularly rich. Every square inch of the main column which carries the stairway is covered with decorative work from the hands of true artists. Quaint little figures, diaper work, decorative columns, canopies, and other features cover it from top to bottom. The details are all fine in character, though there is always the suggestion of quaintness and almost superabundant brilliancy which is a part of the art of Brittany. In general conception and detail it is a distinct art. In general form it may be not unlike that of other sections, yet in the handling of its detail

FIG. 217.—AT NANTES.
(Sketch by A. Robida.)

there is an originality and character which stamp it as belonging to this people. An illustration of this character is particularly valuable, in that it shows the opportunity of new and individual handling of ordinary motives. The artistic methods of Brittany are quite distinct from those of other sections of the world, but structurally and practically their work fulfils the same general conditions as that belonging elsewhere.

Fig. 216 shows the detail of the upper section of this stairway, and gives a better idea of the richness and interesting character of the detail.

If one could see the costume of the old lady who is looking over the rail, in all its color and artistic handling, one would note that the people of the house are not inharmonious with the artistic character of their surroundings.

Fig. 217 shows the landing of a stairway at Nantes.

A German stairway is shown in Fig. 218. There is an originality of plan and general arrangement in this stairway which makes it particularly interesting. Obviously the plan is made to serve an artistic purpose. One can readily understand how a plan of this general character could be handled in a very picturesque and satisfactory manner in one of the halls of our own time.

The stairways mentioned, as well as the one in Fig. 219, are spiral. This arrangement is deservedly not popular with us. While in the instances here set forth they are handled in a beautiful and picturesque manner, it is true that they are more tiresome of ascent and generally more inconvenient than the square stairways with landings, which are more common with us. However, these illustrations are given to indicate the decorative impulse which controlled their designing and execution. The genius and artistic instincts which designed the stairs we have pictured would not be discouraged by any of

Fig. 218.
(From Lachner's "Holzarchitectur.")

FIG. 219. — IN THE MUSEUM AT AMSTERDAM.

the practical problems of our own times. The practical side of a problem does not hamper or even embarrass the true artist.

In the American homes of a few years ago the front stairway was a rather prosaic feature. The old side-hall plan gave little variety, and there was very little incentive for serious effort in decoration. But with the thought which gave the halls a different form, our people, without recourse to foreign examples, saw the decorative possibilities of a stairway. We have them in many forms, often picturesque in outline, though rarely fine or artistic in detail; always with enough labor and material to make them beautiful, but rarely so applied as to accomplish this result. We seldom see a really beautiful stairway. With the ambition which we see on every hand, and with our beautiful material, it needs only the artist to make a great success in this way.

In plans shown on page 102, the stairs are at the end of a hall with wide windows. The broad landings and picturesque railings are between the light and the one who enters the hall from the front. It is expected that landings of this character will be at times partially filled with palms and other decorative plants.

If forms are carefully studied with reference to simplicity, or if the panels of wood are not disturbed by crude incisings or mouldings, the effect can never be bad. The simpler railings and features shown in this book depend largely upon the general plan, the richness of the grain of the wood, and other available qualities, to give them interest. Most stairways are made ugly by an ignorant effort to make them interesting. If one is without natural good-taste or education in design, industry and ambition count for nothing.

The Old Colonial stairways were nearly always beautiful. They were developed in a careful, slow, conscientious way by men well acquainted with the old classic forms. This sort of knowledge was readily conveyed, for the reason that it is easily set forth in text-books. The slow manner in which building was done during Old Colonial times gave opportunity for a serious preparation and study which enabled our earlier architects to do the most artistic building our country has known.

An iron stairway, or one whose railing is of decorative iron-work, hardly appears in place in a modern dwelling. Much beautiful decorative iron-work has been done for the commercial structures of our great cities; there the character of the requirement is so manifest and the general construction of the building so well known that iron appears altogether pertinent. But in a modern dwelling, with draperies, woodwork, pictures, furniture, and other decorative material of a similar character, iron-work, however fine and delicate, appears out of place as the decorative parts of a stairway.

One finds iron and bronze properly used in many of the great châteaux and other monumental structures of the Middle Ages, but there the surroundings are quite different, and stone, iron, and other materials of this general character quite in place.

MATERIALS AND DETAILS. — *Continued*.

CHAPTER XXV.

FURNITURE. — ARCHITECT'S DESIGNS. — SIDEBOARDS. — BOOKCASES. — SEATS. — LOUNGES. — SCREENS. — GRILLES.

THERE are certain furnishings or fittings which it is properly the architect's function to design. We have said that each room should be considered as a whole, having its proper relation to the other rooms of the house. There is very little successful collaboration in painting pictures. It is not to be expected that an architect can partially design a room to be finished by some one else, without the room suffering. Bearing in mind that the room is a picture, its finishing should be controlled and directed by one artist. Of course he may be aided by other artists, carvers, decorators, workers in stained glass, and others. It is not always necessary that chairs, tables, or portable fittings of this character should be designed by the architect. Much good work is now done in this line, and chairs and tables are being made which rely largely on the natural wood and natural forms for their interest. Work of this kind is relatively neutral in character, and for that reason is, within certain limits, suited to any location. In the case of upholstered furniture it is largely a question of the selection of color and pattern. Furniture which shows nothing but the upholstering is nearly always interesting in form because it is natural. Architects very seldom design this class of furniture, though it is sometimes done. In matters of color the artist should always be consulted. During the Middle Ages the architect was responsible for everything which went into a building. All of the fittings, furnishings, and even the hangings and embroideries on the walls were designed by him. There is no more harmonious work than that of that time.

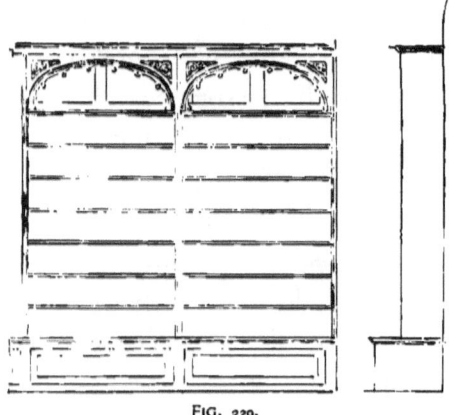

FIG. 220.

There are, however, certain fittings which properly belong to the architect to design: sideboards, buffets, mantels, bookcases, seats, lounges, screens or grilles, portfolios, and furniture for special places. The sideboard is as much a part of the design of the dining-room as the doors or windows, and it is quite absurd to expect that by any accident an architect working in his office and a furniture designer in a distant city each should design something altogether harmonious. One would hardly ask an architect to design a door-casing and then ask some special worker to design the door, yet in a measure this sort of thing is done when the sideboard or the mantel is selected from the work of some special designer who has no knowledge of the room in which it is to be placed. A sideboard or a mantel is a natural part of the room, and the composition cannot succeed unless they are developed by one artist. The designs given for this kind of furniture are for special places, and they belong to the houses and the people for whom they were made, and naturally cannot be a successful part of another house designed by another architect.

Bookshelves nearly always belong to a special place, and even were artistic considerations entirely disregarded it would be absurd to expect to buy bookshelves which would fit a given space where books may be placed. One seldom sees what is generally known as bookcases in a modern house. There may be a few books with fine bindings and other rare qualities which should be protected by glass. But ordinary books look better and

FIG. 221.

are altogether more serviceable when they are placed in shelves in a way to be readily reached without the obstruction of glass doors or other barriers. The design indicated by Fig. 220 is of a typical plain section of

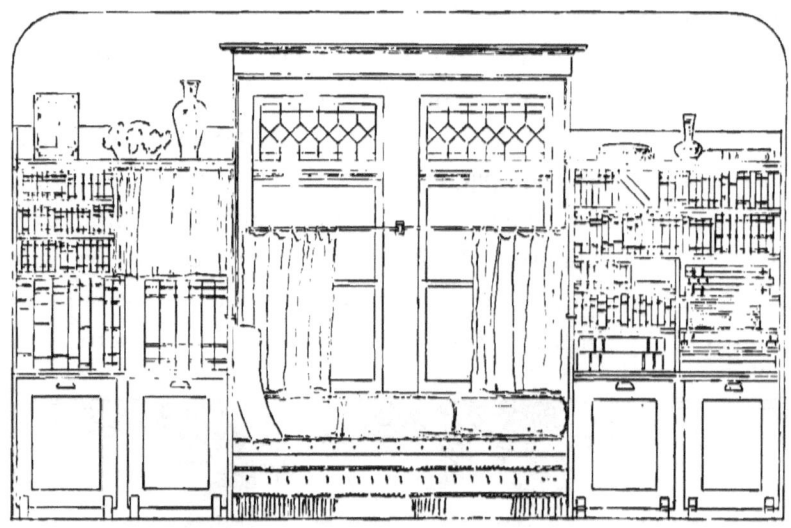

Fig. 222.

bookshelves which form the general model of all fittings of this kind furnished by the writer. It is to be observed that the lower section projects a little beyond the shelves proper. This admits of a little space on which to place a book temporarily. The lower section has panel doors which are hinged from below. In this part may be placed unbound magazines, catalogues, unpresentable books, and other material of like character. The best fitting to use for supporting shelves is a small metallic plug especially made for that purpose. With this arrangement the shelf may be placed at varying heights without taking out the plug. It may be merely turned, and by its eccentricity the shelf may be raised or lowered within certain narrow limits at will. We all know that it is desirable to be able to make finer adjustments of shelves than is usual with the ordinary wood-ratchet contrivance. Portfolios or cases may be placed after the manner shown by Fig. 222. The portfolio can be drawn out into the room in a way to afford a convenient support for other lighter portfolios and documents of various kinds. Very few buildings of moderate cost are provided with proper arrangements of this character.

FIG. 223. — A COSEY SEAT.

Fig. 221 indicates an arrangement of book-shelves, and a mantel which was used in a remodelled house. The mantel is of brick, the woodwork on either side of cherry stained a reddish brown, the various details of which are of the same character as the other woodwork provided for the room. This arrangement provides for a proper display of bric-a-brac and a limited number of books.

Books are the nicest thing about bookshelves, which therefore should not be very elaborate. There is no reason why any large amount of woodwork should be shown in this kind of shelving. It is often well to have a little space above for the display of interesting objects which one likes to have in a sitting-room or library.

The arrangement of books around a doorway (Fig.

FIG. 221.—A SEAT AT PIERREFONDS.

115) is given by way of suggestion as to what may be done. Another instance is given in Fig. 222, which shows bookshelves on each side of a window, and a seat between. A very enticing arrangement of seat and bookshelves is shown by Fig. 223.

Special locations suggest their own interpretation for

uses of this kind, and, as said before, all should be developed by the architect, so that the little mouldings, the touches of finish, the varnishing and all should harmonize properly with the other structural parts of the building.

The varying forms of window-seats, seats in connection with mantels, bookcases, those in niches, etc., suggest themselves in the development of a house plan. At times one finds a space or corner such as shown in Fig. 223. Again it is like that shown at the end of the dining-room on page 196. As a rule, wherever one finds wall space he may find a place for a seat. While one may not care to avail himself of

FIG. 225 — A COUCH

these great opportunities, there is yet the chance. I give here one of the seats from the banquet hall of the Château of Pierrefonds (Fig. 224). These are in niches and arranged along the wall of a great room in which there are splendid fireplaces, tall wainscoting, an oak ceiling with carved beams, richly decorated walls hung with armor, and beautiful windows hung with interesting draperies. The arrangement of such seats is properly a problem for the architect.

In the furnishing stores one sees woodwork which is sold under varying names — spindle-work, fret-work, etc. One can even buy this material by the foot, have it sawed off in the proper length, take it home and nail it up in his doorway to form a part of the confusion which inevitably follows this way of furnishing a house. In one of Fortuny's pictures, "La Vicaria," is a most magnificent screen.

FIG. 226. — RECESSED WINDOWS.

It is of wood, and forms an important part of a great painting. This screen is the work of an accomplished artist. The seat at the right of the room, the shelves by the screen, the woodwork of the balcony, and the door finished with nailheads are all worthy of special attention.

On page 195 the suggestion offered by this picture was used in separating a little music-room from a large hall. There is an idea in connection with the general design of the screen shown in Fortuny's picture. Nearly all work of this kind is made to fit into an opening and reach to the top of it. What a pleasant relief is this suggestion, where the screen is set inside the opening and does not reach the ceiling! As shown in Fig. 139, I have undertaken to illustrate how this idea may be carried out in a modern house.

On pages 147 and 149 are shown designs of grille work for the upper section of doorways. Work of this character requires very careful designing, and, to be permanent, very careful execution. In order to give it the necessary strength there is a temptation to make it heavy. In this instance the woodwork is hardly three-quarters of an inch thick, yet in its construction it comprises six thicknesses of veneer. This work is of the same general character as that in use during the latter part of the fifteenth century in the middle of France.

The screen shown in Fig. 148 which separates the hall in the same structure, is made up largely of forms which belong to the Byzantine architecture of the sixth century.

Fig. 225 shows the picture of a couch in a living-room. Fig. 226 shows a wainscoted room with a low ceiling and high recessed windows with flower-pots.

MATERIALS AND DETAILS. — *Continued.*

CHAPTER XXVI.

WALLS AND CEILINGS. — PUBLIC JUDGMENT. — THE MISLEADING INFLUENCE OF LARGE EXPENDITURES. — PERMANENCY IN DECORATION. — WALL PAPERS. — FRESCO. — JUTE. — DENIM. — SILK.

THE interior decorator has had much to do with making our houses ugly. The pretentious way in which he has done this work and the prices which he has been paid for it have debased public taste. When one pays eight or ten times as much for the so-called decorative work in a room as it is worth, one wishes to think it artistic. The impression is conveyed to the visitor and the neighbor that it ought to be good, that the price is high enough, and that it was done by one well versed in such matters. While people try to admire what is expensive, they cannot be sincere unless there is real merit in what is brought before them.

One is told that an ambitious merchant has paid a thousand dollars for the decoration of a room. Unquestionably this sum of money might do something worthy, but if its expenditure is directed by one without proper knowledge the result will be the worse of the unusual effort. The neighbors go with wondering eyes to see the thousand-dollar room. The colors are inharmonious, the decorative forms are many and bad. Nevertheless they try to admire. They account for their own lack of positive enjoyment in looking at this room on the ground that they are unable to appreciate it. They suppose that it is all right, but they cannot understand it.

The trouble is not with the observers, but with the supposed decorator. Nothing elevates public taste so much as good art. Nothing degrades it so much as pretentious bad art. People wish to admire, and will try to admire, the offensive. Their first judgment is formed upon the representations made to them, and as they cannot really admire the pretentious ugly thing they account for it by deprecating their want of knowledge and taste in such matters. If not that, they blindly pretend to admire and thus deceive themselves. Through such a process there is cultivated a low kind of admiration for ugly things. This kind of taste is fostered by the decorator who says that floral designs are in style this year, or that large figures are the proper thing at this time, and in that way leads the public from one form of ugliness to another.

People are easily impressed by this argument. For some reason many dislike to be out of style. There is nothing which will so quickly dispel false ideas of this kind as a few well-decorated rooms. Most great work is relatively simple, and it is not what the merchant decorator wishes to see perpetuated. There is no change in the style of things that are in good taste. A truly beautiful room is always in style. Good taste is perennial. One who follows the mode wishes to have a different kind of ugliness every few years, without knowing exactly why he tires of it; but one who has a really beautiful room does not care to change. Eight or ten years may pass without a thought of alteration. If the material used is permanent enough in character, there is no reason why it should not be a source of satisfaction and pleasure for ten, twenty, or thirty years. No one can be so brutal as to wish to change really beautiful walls every few years. If they are really ugly, there is some justification in change.

It is only a few years since it became really easy to have good work done in wall papers. A few concerns have employed real artists, and where their products have been used it is not uncommon to see rooms which have remained in the same condition for eight or ten years. It was not easy, however, to get good figured papers previous to that time. Ingrain papers were a great relief from the ugly figured designs of former times. With the introduction of other plain papers of any tint, it became a very easy matter to secure interesting effects at a moderate cost. One is now able to get very beautiful figured ingrains and pulp papers. The art schools of the country and the better artistic spirit have produced designers who do a very high grade of work for wall-paper manufacturers, so that very beautiful rooms can be obtained at a very low cost. Good plain and figured papers are before the people, and if it only become a universal custom to buy really good things a great work will be done at once. A great majority of the people have not the courage to depart from general methods, even where everything is very wrong. The few courageous people who do the right thing because it is right must in time affect the multitude. There can be no question as to the ultimate effect of a good example.

Beautiful and permanent wall-patterns in stamped paper are now produced, so that the medium for securing good results cannot be said to be wanting. If the people will only disregard the talk of the salesman about what is or is not the style, and select what is artistic and beautiful for its own sake, the period of ugliness will soon pass.

There is a tendency to undervalue the character of work which may be done with wall-papers, because so much bad work has been done. It is looked upon by many people as a common and ordinary material. But the question is merely one of design, color, and drawing. As we now have all this in wall-papers, we may be certain of securing satisfactory results. A paper surface is a good one, from both an artistic and a practical standpoint, and with the improvement in design there can be no question about its extended use.

With the improvements in paper-making, there has been developed material which lends itself readily to the production of various surfaces. The artistic quality of any product is dependent entirely upon the one who gives color and design to it. There are papers with stamped surfaces, in which there is no effort at imitation of other materials, which are very satisfactory. Paper lends itself to an infinite variety of uses, one of the most satisfactory being the production of designs in low relief. The artistic value of such material depends upon its having been shaped and colored by an artist.

Cloth surfaces of silk, cotton, jute, and other material

of that character are used for wall decoration. Very interesting results are secured through such mediums. Light cotton prints, cretonnes, and other materials whose name and variety are legion, are used in this way. There can be no question as to the artistic value of these substances. However, some are naturally absorbent, and are usually applied by tacking to the wall surfaces. Their sanitary value may be seriously questioned. Heavier cotton fabrics, such as jute and denim, are often used and applied in the same manner as wall-papers.

Jute has a fine natural color,—a rich tawny brown, — and is capable of being stencilled or otherwise decorated in colors in a way to form a permanent and very beautiful covering. Denim admits of the same character of treatment. It has an individuality of color and surface which is very satisfactory. It is much better that material of this character should be left entirely plain rather than be decorated by an uncertain hand. The great value of ingrain paper and other materials having a plain surface is that there is less danger of failure when handled by ordinary ability. It is lamentably true, however, that there are very few people but have the assurance to undertake the most difficult of problems in the decoration of wall-surfaces; viz., the selection of material, color, and design. While the color sense is lacking in most people, few indeed are aware of their infirmity; and the number who can judge the drawing of figure designs is even less.

It is hardly possible to enumerate the various materials and the character of surface which may be applied to walls as a means of decoration. The general principle in the formation of successful designs can never vary. The mediums may increase in number, but the general principle covering their use must remain the same.

The frescoer has perpetuated much ugliness in house decoration. He has been employed by the more ambitious house-owners, and has been pretentious and dogmatic. He has received liberal compensation, very much more than his services have been worth. He has covered walls with twining floral designs; he has pictured birds, or made bad copies in form and color from the rococo period; and altogether has done a great deal of very ugly work. Not many years ago a few ambitious and fairly well-trained young men in the East began to do decorative work of this character in a better way. They were very successful, and their work commanded such good prices that it was soon out of the reach of the masses. But from the demand for this sort of thing there has developed a larger supply, so that we now find in larger cities a few decorators who are doing sincere work. This is educational. In time we may expect to see satisfactory general results.

One must naturally be very careful in dealing with many mere professionals. There are many who are unworthy, and who make up in skill as salesmen what they lack in artistic ability. They soon learn to ape the manners and sincere expressions of real, earnest, well-educated designers, but their work is always a failure in execution. Another source of danger comes from merchants who employ artists to make meritorious water-color sketches to show to the patrons, but who execute these designs through a very common grade of workmen.

It is true, however, that the number of conscientious and well-educated decorators is on the increase. Within a few years we shall have a better education and a large number of successful workers in this branch of art. Under the difficulties of recent years, it is not to be wondered at that people have done very little that is permanent in house decoration. Now that a better grade of ability has entered this field, there is no reason why we should not think seriously of using permanent materials in the decoration of our homes. The establishment of art schools in nearly all of the cities of our country leads to this state of affairs. An artist can express himself quite as well on a plastered wall-surface as upon a few square feet of canvas. The training which a young man or a young woman receives in an art school in drawing from casts and figures, and the general training of the mind in the appreciation of things artistic, are exactly what is necessary to develop house decorators with the artistic quality. While it may be a worthy ambition to become a great painter, it is well to bear in mind that such a one must have more than the ability to color and draw well. He must combine with these the ability of a great poet or philosopher in order to paint pictures to reach the human emotion. Not all can do this. Many who undertake it would do much better and better play their part in the world by learning to appreciate the work of the great painters, and, through the mental training thus afforded and the manual dexterity gained in the art school, by intelligently decorating buildings.

The thought of permanency in decorative wall-surfaces may not at once impress any large number of people. So much of the work having been done unsatisfactorily in the past, there is no reason why it should be permanent. It is better that it should be temporary. But the work of an artist may well be permanent. There is no reason why it should not last as long as the building. There is no reason why it should not be done with material which is as lasting as the other parts of the structure. It is well for a healthy mind to live in a well-decorated building without change. If all decoration were beautiful and artistic, and so became a part of our lives, it would be sad indeed if it should be taken from us. It should be to us as our pictures, our books and furniture.

Many good houses are planned by conscientious architects, and turned over to the owners with their white or gray walls to be turned into chambers of horrors. Inharmonious colors and bad designs run riot. No amount of business success will in itself render a business man capable of decorating his own house. No amount of social tact or home-loving qualities will enable a woman to do the same thing. It is a mere question of artistic training. Not every housekeeper or successful business man can decorate his own house.

MATERIALS AND DETAILS. — *Continued.*

CHAPTER XXVII.

MATERIALS. — KINDS OF WOOD. — MOULDINGS IN WOOD. — PLAIN SURFACES. — THE WOOD SCREEN AT AMIENS. — STAINING OF WOOD. — WOOD FINISHES. — WOOD FLOORS. — WORKMANSHIP.

A MATERIAL which may add so much to the appearance of the interior of a house as wood should have serious consideration. Much more could be done with this material than is usually accomplished. Much beautiful wood is absolutely ruined, either from the forms which are given it, or the way in which it is finished by the varnishers after it has been placed in position. It is for this reason that people do not realize the artistic value of plain surfaces of wood. We commonly find doors, casings, and other woodwork loaded with clumsy and ugly mouldings. I am not pleading for absolutely plain surfaces, nor for the doing away with mouldings or other decorative forms. Far from it. Properly formed mouldings and decorative surfaces in the right place are necessary; but they must be suited in character to the material used, and must be a means to an end rather than the end itself. Forms may be devised either to emphasize and accentuate the natural beauty of moulded and decorated surfaces, or to bring natural plain surfaces into their proper relation. We often find on the interior of a house decorative forms and mouldings which are suitable only to exterior stonework. It is true that certain stone forms are suited to wood, but not all are.

In designing the interior woodwork, the architect has the natural beauty of the wood to help him. He can handle it in a characteristic way. While stone is a pleasing material, there is not that help from its texture or color which comes from wood. The beauty of stonework is largely in the distinctive forms which are given it. Wood has its own natural beauty, which may be rendered yet more interesting by the forms given, if they are considered with respect to the beauty of the material. But no matter how interesting in themselves, if the forms be unsuited to execution in wood they will destroy the beauty of the material itself.

In following these principles one must not be too sweeping. A Doric column is essentially stone construction, yet the mouldings thereof and the form of the column itself, particularly if the Roman channels be not used, are well adapted to showing the grain of the wood. What could be more beautiful than a Doric column of maple on the interior of a building, and finished in its natural color? The mouldings at the base, the plain intervening surface from the base to the cap, the cap mouldings, the shadows coming thereon from the reflected light of a room, are very beautiful. The Ionic cap is not a particularly interesting form for wood construction. A Corinthian capital is not adapted to wood, particularly to heavily grained woods, such as oak or sycamore. Altogether, we find that the more elaborately decorated orders are less satisfactory in wood than the plainer ones. Plain Greek mouldings always look particularly well in wood. The curves are soft and gentle, the lines and recesses sharply accentuated. One gets the variety of form, the decorative interest, without sacrificing the natural beauty of the material. Reference to some of the Greek mouldings specifically mentioned in this volume will illustrate this point.

A great deal of money is often spent upon mere workmanship, without artistic direction, when if one were somewhat more particular in the selection of material, and kept the surfaces much plainer, the cost would be much less and the artistic result much more satisfactory. It is not uncommon to see pine or poplar used to save a little money in the cost of material, nor is it at all uncommon to see enough money expended in mere workmanship to have permitted the use of plain mahogany instead of the elaborate pine or poplar. There is no artistic gain in making a very elaborate mahogany door.

Pine can be made to look very beautiful, but it can hardly be kept so permanently. White pine finishes well, but does not stand the ravages of time. It turns yellow, soon becomes marred, and in a few years is far from beautiful. This is also true of poplar. Poplar stains well, but is too soft for use. The ideal material for finish is one which resists all ordinary use and continues to improve with age. This is true of almost any hard wood, particularly oak. In the structures of the Old World, one finds splendid decorations in that material which are beautiful to-day after several hundred years of use. There is no more beautiful piece of woodwork in existence than the choir stalls in the cathedral at Amiens, which were executed in the fourteenth century. The work itself is quite as substantial as it was on the day it was finished. During the Revolution the fleurs-de-lis were scraped from the panels, but otherwise it is unharmed. One cannot make a mistake in using oak for the finish of his house. Just now quartered white oak is a relatively inexpensive material, and quartered red oak even less costly. The difference in the cost of finishing a room in oak or in pine or poplar is not great and is in no way commensurate with the difference in value. The soft wood depreciates greatly in a few years, while oak takes to itself a color and richness which render it more valuable with age. Soft wood depreciates and hard wood appreciates in value.

Maple has been mentioned. This is a very pleasing material for a dainty chamber, or a light, delicate reception-room. Its great beauty, however, is its light-ivory color, which it cannot be expected to keep much beyond eight or ten years. It has been a habit to stain maple, yet birch is more interesting when treated in this way. Birch is a hard wood and admirably suited to staining. It is also a good material to enamel-finish. One frequently sees poplar, or whitewood, which is the Eastern name for this material, enamelled or even ebonized. But it should not be so used, for the reason already given — the body is not hard enough. Enamel and ivory finishes should be applied to hard woods, to birch or maple. Cypress is a reasonably hard wood well suited for such uses, though its natural grain should be obliterated as nearly as possible. It should be used only for some sort of paint finish. The natural grain of cypress is as violently ugly and coarse for interior finish as yellow pine.

No wood presents a greater variety of surface than quarter-sawed sycamore. It finishes beautifully, but becomes tiresome if much of it is used. The first impression made by a room finished with this material is very satisfactory. On one hand there are surfaces which rival bird's-eye maple; near it others with the variety of quartered oak; again other bits quite as pronounced and unusual, as curly birch; and still others so peculiar in character that one finds no basis for comparison. People are usually very enthusiastic when they first see sycamore, but they soon find it too aggressive in beauty, and weary of it.

Cherry has a splendid hard texture capable of being stained and beautifully finished, but, excepting for its variety, it is not as interesting as some other hard woods; generally it does not improve with age. It is now almost as expensive as mahogany.

One might further consider this subject of woods, but without particular gain. There is a tendency on the part of many people to seek great variety in the finishing of a house, to have a different wood in every room if possible. There is certainly no artistic advantage in such an arrangement. While there is no objection to a certain amount of variety, the restless spirit which does away with all repose is certainly obnoxious.

The problem of the hard-wood floor was a good many years in reaching a satisfactory solution. There can be no question as to the sanitary value of such a floor. Reasoning from this standpoint we would say, Place them all over a house, although they must be largely covered with rugs. Rugs have a great advantage over carpets, because they may be readily cleaned. But it is the constructive problem which has been troublesome. Several years ago the writer put down a great many solid hard-wood floors. In order to keep them clean they had to be put down after all other woodwork had been placed in position, and all work of all kinds about the building had been finished. Then only could they be laid and finished. If the floors were laid after the other work remained to be done about the building, they were sure to be soiled; again, the intense heat of the furnace caused the joints to open, maybe but slightly, but enough to make them unsatisfactory.

A much better way to get a hard-wood floor is to first put down a covering of narrow yellow-pine flooring, and after the house has been finished to lay one of the hard-wood floors made by specialists under the varying names of wood carpeting, parquetry flooring, veneer floors, etc. If one uses the plain patterns of quartered oak the cost is not great for a first-class floor, and the result is much more satisfactory than by using elaborate designs and different kinds of woods. A number of years ago when I said that I did not believe a hard-wood floor could be made satisfactory. I was referred to the splendid hard-wood floors in France and Germany. As a matter of fact, one sees many well-cared-for floors in these countries, but few or none which are constructively satisfactory if laid in solid strips. The floor of the Louvre receives excellent attention, but has cracks into which one can readily drop a match. This floor and the one at Versailles have probably done quite as much to impress Americans with the beauty of hard-wood floors as any others; yet it is the manner in which they are cared for, rather than the floors themselves, which commands attention. This is also true of the floors in most private structures in Europe. Fig. 195 shows the usual pattern of the French floors.

Various preparations for floor finishing are now made which are quite satisfactory, and which wear for a long time without renewing.

After one has done all that one may in the preparation of drawings and in the selection of materials, there is yet the controlling influence of workmanship. Nothing will atone for its deficiency. We speak of beautiful surfaces of wood, yet if they be not carefully executed their very plainness is uninteresting. Good design is essential, but poor workmanship may destroy all.

All good hard-wood work is veneered for constructive reasons. A door is almost certain to warp if it does not have a soft-wood laminated core. All of the large plain surfaces of beautiful wood must be backed up with laminated soft wood in order to preserve it in its proper form and appearance. These are matters of constructive detail which have been carefully considered in another volume. It is sufficient to say, however, that the standard of workmanship is improving with years, and as time moves on it is yet easier to secure satisfactory execution.

MATERIALS AND DETAILS.

The finishing of hard woods by the varnisher is more successful than it was a few years ago. The manufacturers of different kinds of wood finish have greatly improved both their products and the manner of marketing them. They now give careful directions and samples illustrating their proper use.

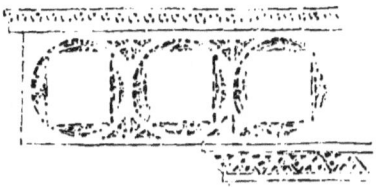

A DECORATIVE MOTIVE.
(Louis H. Sullivan, Architect.)

MATERIALS AND DETAILS. — *Continued.*

CHAPTER XXVIII.[1]

THE ARTIST BLACKSMITH. — EARLY WORK. — HINGES. — LOCKS. — NAIL-HEADS.

THE ancient blacksmith was an artist. He expressed the art feeling in iron over the anvil, by means

FIG. 227. — A TWELFTH-CENTURY GRILLE.

---

[1] All the illustrations in this chapter are from Viollet-le-Duc's " Dict. de l'Architecture."

of his hammer and the stamping-tool, in the same
general way that the sculptor expresses himself in clay

FIG. 228. — OF THE THIRTEENTH CENTURY.

or stone. To him iron was a medium, and his technique
was as serious a question as is that of the painter.
Compare the hinges of the great oak doors of Notre
Dame of Paris with the work of the other artists who
were employed on the same edifice, and we may judge
of the seriousness of the effort. This was the work of
the thirteenth century.

Fig. 227 is a grille — an iron screen — of the early
twelfth century. The delicate work around the frame is
done with the pointed tool — a kind of punch. To

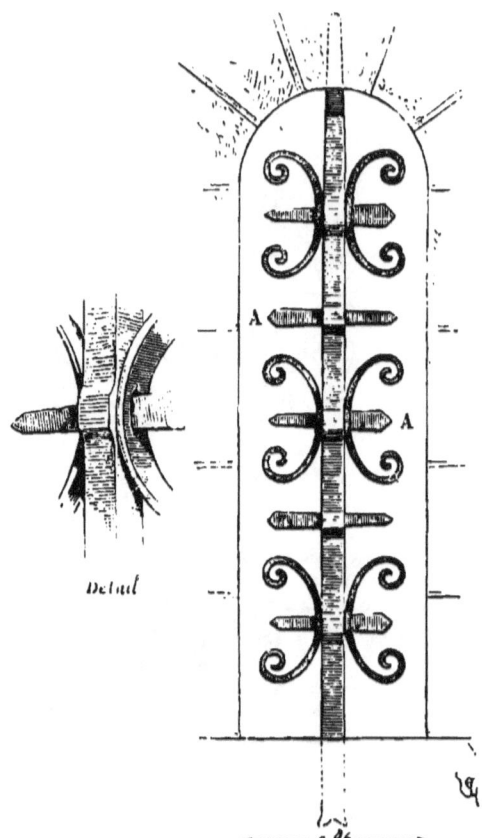

Detail

FIG. 229.—COMMON IN THE TWELFTH CENTURY.

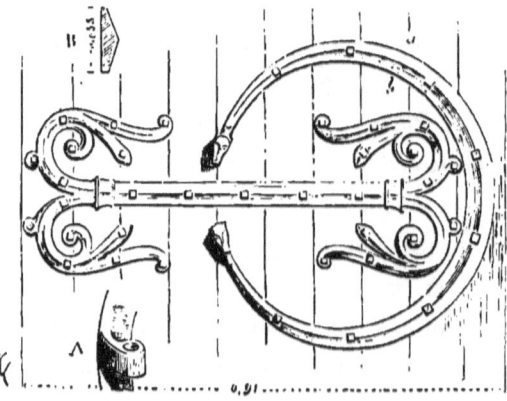

FIG. 230.—THE ELEVENTH CENTURY.

enrich the other surfaces, pointing of the same general type covers the face of all of the tendrils and the upright muntins.

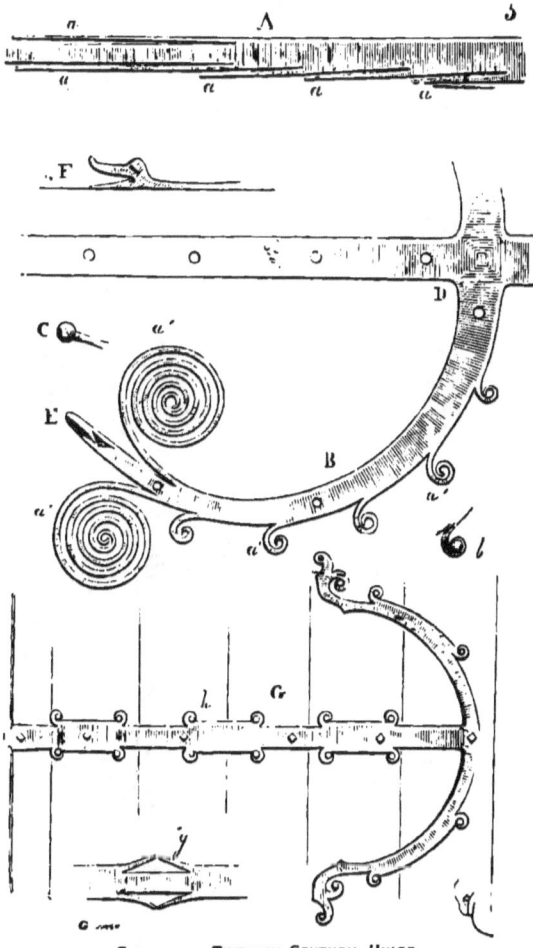

FIG. 231. — TWELFTH-CENTURY HINGE.

Fig. 228 is a grille of the thirteenth century, from St. Denis, near Paris. The marks of the various tools for enriching the surfaces are plainly shown in this design. Matrixes of various designs were used to give to the surfaces the various forms shown.

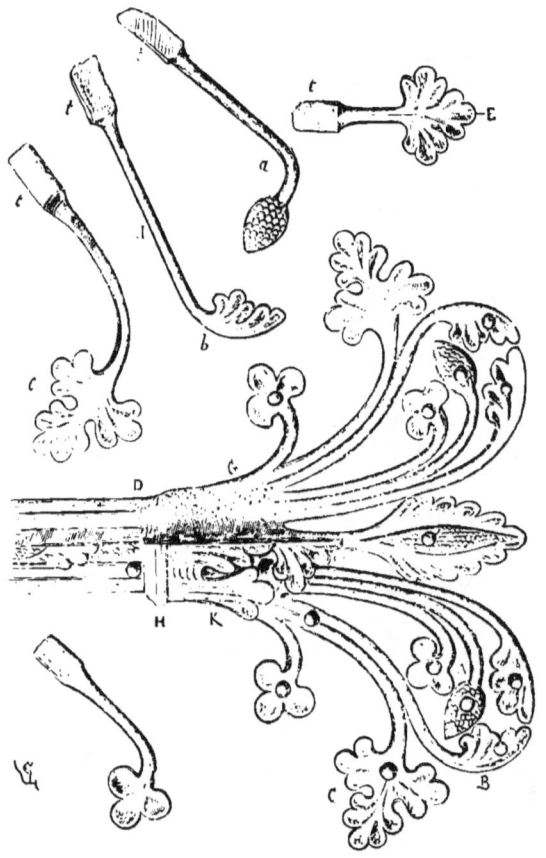

FIG. 232. — FROM NOTRE DAME, OF PARIS, THIRTEENTH CENTURY.

Fig. 229 is a form used in windows, and was very common in France during the twelfth century. It is to be remembered that during that period the merchant forms of the iron, as we would speak of them, were given by hand with hammers and on an anvil, rather than by rolls and the mechanical methods known to our

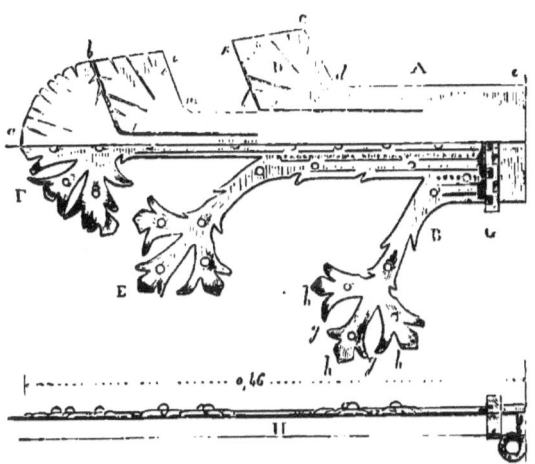

FIG. 233.—OF THE FOURTEENTH CENTURY.

own times. We may know of all of the advantages of machinery, and yet realize that because of the convenient

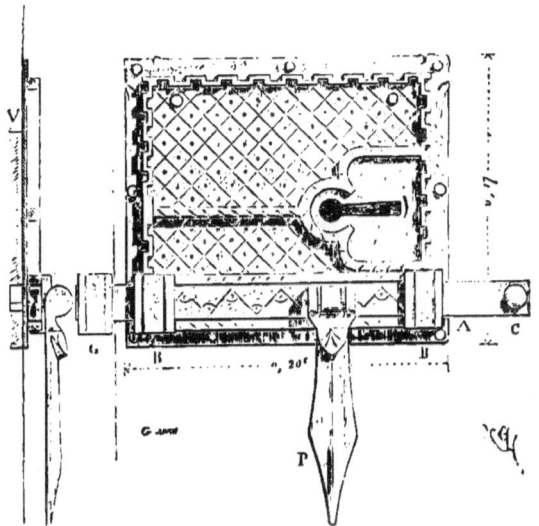

FIG. 234.—A TWELFTH-CENTURY LOCK.

methods of these times for producing the forms by machinery our men have not only lost their manual

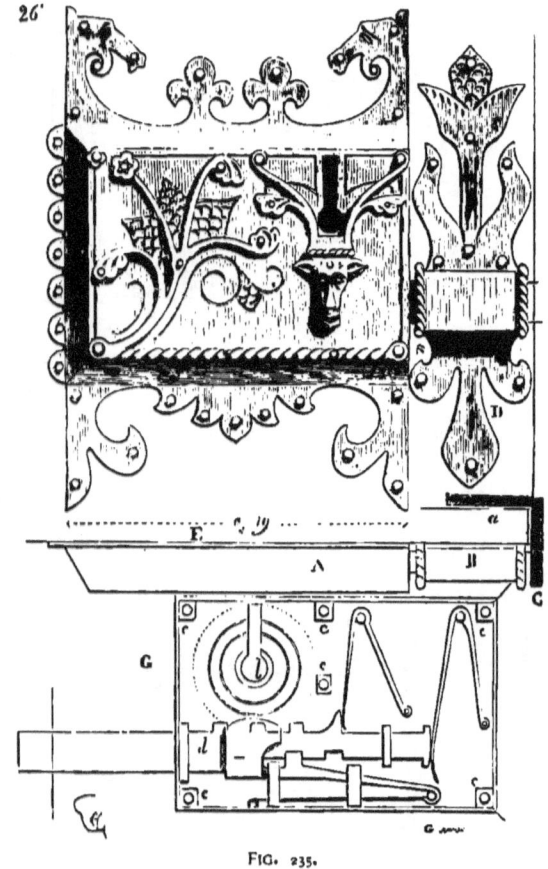

FIG. 235.

dexterity, but also the spirit of the artist. There can be no question about this.

Hinges reached their perfection between the twelfth and thirteenth centuries. The metal work was used not alone to hang the door, but as well to hold the door itself together. Fig. 230 is a hinge of the eleventh century. Fig. 231 shows forms and methods of construction peculiar to the twelfth century. Figures with the little tendrils cut from the sheet of metal are shown in their curled decorative forms in this illustration. The same general method is indicated by Fig. 233. Fig. 232 indicates some

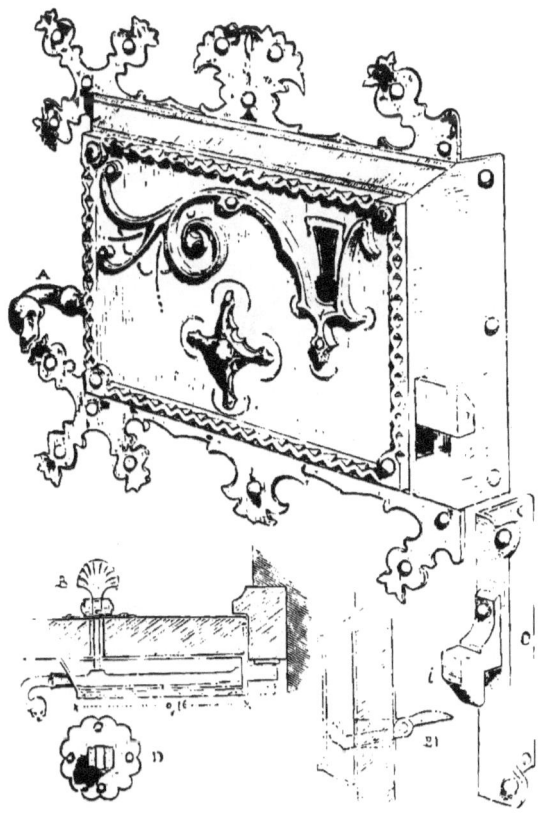

FIG. 216.

of the details from the hinges on the great doors at Notre Dame. Fig. 233 is a hinge formation peculiar to the fourteenth century. The rosettes are stamped and

326   BEAUTIFUL HOUSES.

FIG. 237. — OF THE FIFTEENTH CENTURY.

welded to the tendrils. There is an opening in the centre through which the nail is driven, thus filling an

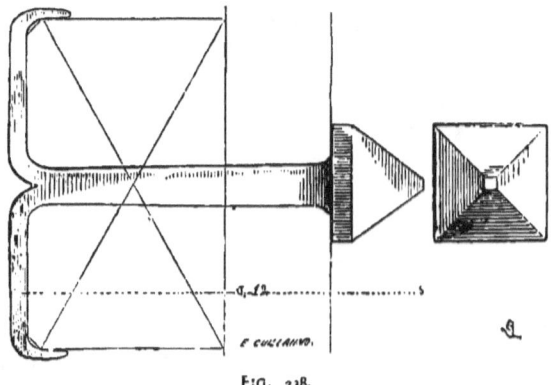

FIG. 238.

important decorative office. These hinges and the door formed one complete composition.

We must satisfy ourselves with a few examples of

FIG. 239.

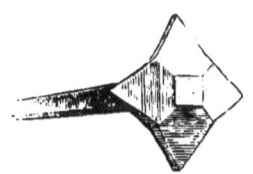

FIG. 240.

locks: Fig. 234, dating from the twelfth century; Fig. 235, from the thirteenth; Fig. 236, from the fourteenth; and the beautiful work, Fig. 237, from the fifteenth century. The abundance of beautiful examples of this kind of work is so large as to be embarrassing when one wants to make a selection.

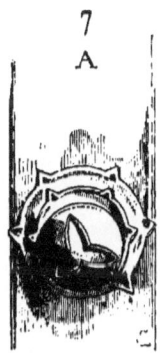

FIG. 241.

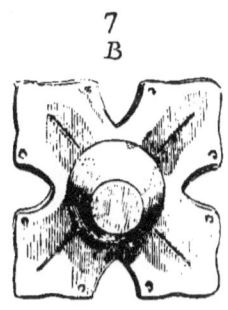

FIG. 242.

Of art expressed in nail-heads I can select but a few examples. — Figs. 238, 239, 240, 241, 242, 243, — which range from the twelfth to the fifteenth centuries, inclusive. They picture their own description.

The forms are all natural. There is no attempt to imitate those not adapted to metal. In the seventeenth and eighteenth centuries there was an effort to make forms which were better adapted to any other material. In that period we find perfect imitations of fruit, leaves, and other natural forms. Decorative stone work was even imitated in iron, but the metal forms were rarely of natural objects.

FIG. 243.

MATERIAL AND DETAILS. — *Continued.*

CHAPTER XXIX.

GLASS. — DECORATIVE USES. — RECENT FAILURES IN GLASS WORK. — ARTISTS' WORK IN GLASS. — SUCCESSFUL USE OF COLOR.

THE history of glass merely shows that the useful has taken decorative prominence. It is certain that the original idea of glass was to let light into the rooms, and windows were placed solely with regard to their usefulness. Now windows are placed so that they will look well, and their useful forms are made decorative. This is the history of all decorative features; their origin was in utility.

The history of glass does not concern us, however. Recent attempts to use glass in a decorative way in our houses have not contributed to the progress of the artistic. In recent years we have passed through various stages of horrors in the use of glass. During the Old Colonial time glass in small sizes was used because it was inexpensive. The sash was divided into small pieces by wooden muntins. Interesting forms were often placed over and around the doors. After a time, however, glass came to be used in larger sizes, and during recent years it has been the ordinary practice to have but a single glass in a sash. The advent of plate glass favored this tendency to large sheets. Plate glass in large sizes came to be a fad, and we even now see great show-windows of this material built into moderate-sized homes.

After the larger-sized sheets had come into general use a relief from the plainness of these large openings was sought. Then the upper sash of windows, the transoms, the doors, and single sashes, were divided in various ways by wooden muntins. Sheets of colored glass were placed in them. More often than otherwise the selections were made by people incapable of combining color successfully. Not infrequently violent reds, blues, greens, and yellows were placed in the same sash. In time it was realized that this was not beautiful, and there was a search for something new in colored glass. There came ribbed glass, opalescent glass, and frosted glass. But all were relatively cheap, and they went into all kinds of houses.

FIG. 244.
(From Viollet-le-Duc's "Dict. de l'Architecture.")

Later came the so-called art glass. The sash was provided with leaded designs; the patterns were quite unusual, very elaborate, filled with many kinds of glass, each under a new name. The coloring was startling, nearly always inharmonious. Large prices were paid for these exaggerations, and owners are known to have estimated the artistic value according to the price per foot. In the course of a relatively short time, however, these ugly things became common, and, being ugly, the people began to suspect them. The manufacturers had something else ready for them. The country was filled with bevelled plate glass. It was leaded in all imaginable shapes, combined with jewels, bulls' eyes, stained glass, and in other ways made quite as expensive as the other designs. This glass had its run, and again there was a suspicion that there was something wrong, and now nothing is wanted but plain sheets. This simply proves the old adage that all of the people cannot be deceived all the time. When they revolt from a thing of this kind they go to an unnecessary extreme. If the stained and leaded glass had been arranged and designed by artists, this reaction would not have taken place, and many homes would have been much more beautiful than they are to-day.

There has never been a reaction from stained-glass designs arranged by true artists.

FIG. 245.
(From Viollet-le-Duc's " Dict. de l'Architecture.")

The greatest known to us are those of the thirteenth and fourteenth centuries. The world can never grow tired of such glass as is found in the Cathedral of Chartres or

Nurenberg. Viollet-le-Duc, a great French architect and archæologist, in speaking of an early experience when a child, during a visit to Notre Dame, in Paris, says: "I was often confided to the care of an old domestic, who led me to walk wherever his fancy dictated. One day he took me to the Church of Notre Dame, carrying me in his arms, for the crowd was great. My attention was attracted to the glass of the south rose-window, through which the rays of the sun penetrated, colored by the most radiant hues. I still seem to see the spot where we were stopped by the crowd. Suddenly the great organ rose into music. To me it was the rose before my eyes which sung. My old guide sought in vain to undeceive me. Under this impression, more and more lively when such panels of glass produced the graver tones and such others uttered the high and piercing ones, I was seized with such terror that it was necessary to take me out."

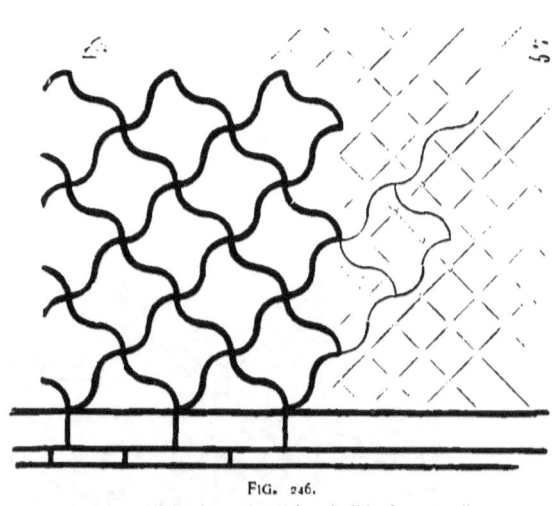

FIG. 246.
(From Viollet-le-Duc's "Dict. de l'Architecture.")

It seems, therefore, that it is not education alone which establishes within us these intimate relations between the various expressions of art. There is an emotional influence about a serious work of art which is never without its effect, and it is a sad thought that where we might have had so much that is beautiful, there is now so much that is ugly, made so from the vulgar desire to do the unusual. Just at this time the majority of people who build are hard to interest in any kind of leaded glass.

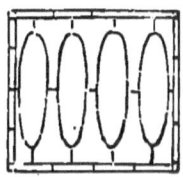

FIG. 247.

Of course this state of things will not last long. The influence of what is really beautiful will make itself felt. and in time we shall have beautiful designs. The Old World affords us many examples of plain leaded glass, really inexpensive and of great artistic value. Several designs of this kind are given in this chapter. We can use plain. clear, double-strength American glass. In small, neat patterns it makes a very interesting window. These same patterns can be carried out in lighter tints of cathedral glass, though not so much can be used in a building as of the leaded sheet-glass. The successful use of colored glass and painted and stained glass depends entirely upon the presence of an artist to control the disposition, coloring, and arrangement of the more elaborate patterns. If one is not skilled in the use of color, or is without some one in whose ability he has full confidence, it is best to use the plain leaded glass, or small quantities of amber and other light tints of cathedral glass in the same leaded forms. One is safer in using only one color, and that in light tints.

FIG. 248.

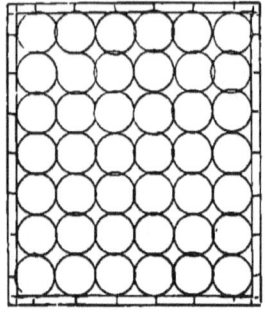

FIG. 249.

The color of the atmosphere in a room may be largely controlled by the proper use of cathedral glass. This is a matter which has not hitherto been properly considered in connection with the planning of a home. There is no one thing which will do more to make a pleasant living-room than the proper arrangement of colored glass. People have been so greatly outraged by what has been done for them with leaded glass that the reaction will be slow.

FIG. 250.

MATERIALS AND DETAILS.— *Concluded.*

CHAPTER XXX.

MACHINERY AND THE ARTS. — SCULPTURE WORK. — MOULDED BRICKS. — COLOR IN BRICK.

FEW of us appreciate the improvement machinery has wrought in the manufacture of brick. This is not alone through their reduction in cost or the improvement of the ordinary form, but as well in the production of artistic forms. While most moulded brick and those of decorative shapes are uninteresting, there are those which are truly beautiful and artistic.

Some of us decry the use of machinery in the production of things artistic. It is affirmed that the artistic which is in the mind can find expression only through the intervention of the tool which is in the hand.

Certain grades of sculptural work and certain kinds of carving will always have to be done by hand. Yet it is true that many decorative motives can be produced and reproduced as well through the agency of the machine as through the agency of the hand. It is the artistic impulse of the mind guiding the hand which produces the result. Why may it not be true that the same artistic impulse and the same mind which guide the hand may also guide the machine?

There are many manufacturers of moulded brick who

are producing by machinery very clumsy and very awkward designs. The mind which guides the machine, the mind which produces the mould from which these clumsy forms are made, is a clumsy, untrained one. The hand which would give expression to the thoughts of that mind is equally clumsy. It certainly multiplies ugliness.

A manufacturer of moulded brick has employed one of the greatest sculptors in this country to make the models for his moulds. What is the result? A reproduction at a moderate cost of beautiful decorative forms. The hand which produces the moulds for the brick never did anything better when working with the modelling tools, or the chisel and hammer. Thousands of these brick are reproduced and sold at a moderate cost, whereas by the method of working directly with the tools, only a few feet of the same decorative form could be produced in the same time, and at a cost infinitely greater.

Some would disassociate the idea of low cost from the artistic. But a form must be artistic because it is beautiful, not because it is expensive.

There are those of us who immediately lose interest in a beautiful form because it is frequently reproduced by a machine. Such an idea is not creditable to any one. The frequency of reproduction of a truly beautiful design does not affect its beauty or its interest. The Greek honeysuckle ornament has been in use for the past twenty-five hundred years, and its repetition has had no effect upon its beauty or the appreciation of it by the public. Certainly its reproduction has been largely through the direct agency of the human hand, but that hand has often been as much of a machine as any device in a machine shop.

FIG. 251. — A SIXTEENTH-CENTURY DOORWAY.

The use of the acanthus leaf comes down to us from the early periods of Greek architecture. It was universal in the Byzantine and the Roman architecture. There were traces of it in the Gothic, and it springs to new life in the Renaissance. Yet its reproduction by the machines called men in the Roman times has never had the effect of lessening its interest in the artistic mind. A form which was reproduced an infinite number of times during the period of the Roman and the Renaissance architecture, and which we see on every hand to-day, is the egg and dart moulding.

Fig. 251 shows a doorway of the sixteenth century. In it the ambitious decorative clayworker will find many forms capable of reproduction through an intelligently constructed and an artistically guided machine of his own creation. If he were to put a clumsy hand to the reproduction of a doorway like this, he would have a clumsy, inartistic result. If he were to give to a highly refined, educated artist the task of arranging and guiding the work of the machine, he would have a correspondingly artistic result. The machine tool is within certain limitations capable of artistic results in the same sense as is the hand tool.

Red has been the usual color of brick because most clay burns red. The improvement in the science of mixing and securing various colors has led many to consider whether or not the use of red brick for the large proportion of our best buildings is not a thing of the past. Buff, gray, and brown brick of various shades are readily obtained, and are used in the large cities almost to the exclusion of red brick. Bricks with clouded and mottled surfaces, and those of differing proportions, are being used. The Roman brick, which is about an inch and a half thick by ten inches long, is very pretty. But there is a tendency among our people to go to extremes in all these matters.

Many attempts have been made in our country to develop patterns with different-colored brick in the wall-surfaces, but as yet this has not been attended with any great success. The contrasts have all been too strong. In France and Holland, where one sees this done to the best advantage, there is only a slight difference between the varying colors, and by that means very pleasing effects are often secured. In recent years the French have undertaken some rather brilliant work with colored brick with indifferent success.

A DECORATIVE MOTIVE.

THE ARCHITECT.

CHAPTER XXXI.

THE ARCHITECT. — THE ARCHITECT AND THE HOUSEWIFE. — BUSINESS AND THE ARTS. — COSTS. — PROPER UNDERSTANDING OF THE CLIENT'S WISHES. — PLENTY OF TIME TO MAKE PLANS.

SAY what one will about the artistic in house-building, the architect must be a business man. What will it cost? is a vital question with the one who would build him a house. The architect must be able to tell what it will cost, and so conduct the business of house-building as to relieve the owner from the care and annoyances which are usually associated with building.

Some years ago the writer prepared a volume upon the severely practical matters pertaining to house-building. It was based on the conception that there is a definite relation between the work of the housekeeper and that of the architect. In it were given plans of houses of various kinds, most of them being of low cost. In connection therewith all of the practical details of the house were minutely considered. The practical side of house-building was so fully considered that little mention was made of the artistic side. In reading the present book one may think that the practical has been neglected. While they have not been especially mentioned, the housekeeper and the architect have been associated in their proper relation in connection with all of the plans which have been presented. The especial mission of this book is to set forth the relation of the artist to house-building. It is "special" along its line, in the same sense that the book on "Convenient Houses" was special in its way.

The work of the domestic architect is very complicated, involving a large amount of special knowledge of the many features of house-building. It is the large amount of detail which has made the practical work of house-building very annoying. But system and orderly methods remove much of this. An architect properly trained to his work can tell the prospective builder exactly what his house will cost before he has created any other obligations than those which are connected with the preparation of plans. The architect cannot tell at the first interview what his client's house will cost, nor can he tell him exactly until the plans are finished. When the client applies to the architect, either in person or by correspondence, neither knows exactly what is wanted. The process of making plans is a method of determining this. The proper work of the architect is in so forming these plans, and developing the specifications, that all matters pertaining to the house are properly represented and set forth by the plans and specifications. Such a thing as an extra charge from a contractor never need occur, unless the client changes his mind with respect to some detail of the building which involves additional labor and material. Except under these conditions, the cost may be known exactly before the client signs his name to one of the building contracts. If the business part of the architect's office is properly organized, it is next to impossible that there be anything forgotten or omitted.

It may be well to indicate how it is possible to think of everything and include everything in connection with the preparation of a set of plans, without chance of error. In a properly organized office, everything which relates to the business of house-building is done from forms, so that in the preparation of specifications, for instance, they are made up from printed sheets which include everything which may go into a house. The parts which do not go into the house under immediate consideration are marked out or crossed off from the form. These sheets are first handed to the owner, who may learn that he is not to have inside shutters, or that there is no plastering in the attic; that the floor in the cellar is not of brick, but of cement; that the roof is covered with shingles, and not with slate. In truth, the client knows everything he is to have, and there is set before him everything he is not to have. There is more chance of failing to omit something than there is of failing to include any detail. This not only gives positive and negative information, but brings everything to the attention of the client. The specifications are made by copying the parts of the printed specifications not omitted. It is only by a method of this kind that mistakes can be avoided. When one trusts to memory or general knowledge or unformulated experience, no matter how great, mistakes and omissions are sure to occur on every piece of work. A plan of this kind means the application of rigid, formal business methods in an architect's office. Through their use many houses have been erected without one dollar being paid for extra work, and at times settlements have actually been made for less than the contract price. This resulted from the omission of certain parts, or from the collection of over-time damages.

One of the most difficult features connected with the business part of the architect's work is the proper understanding of the real needs and requirements of the client. Usually he does not go to his architect until a short time before he is in need of plans. This necessitates haste. The drawings may be never so well made, and the specifications never so accurately formed, yet the client may not have had sufficient time properly to develop his actual requirements. It would be much better if every set of plans should be under consideration six months or a year before they are finally finished. Any conscientious architect would be pleased to carry forward plans in this way. Not that he would be working on them all the time, but he would have that time to study the definite requirements of his client.

First there would be a set of sketches submitted. Instead of taking them home and looking them over in the evening, as so frequently happens, there could be a week or two for serious consideration on the part of the client. These sketches, being returned to the office for amendment, could be leisurely worked over again, and again returned, and so on many times. In the end there would be what every serious worker so much covets — general satisfaction with all details. A plan of operations of this kind is an education to the architect as to the definite requirements of the owner, and an education to the client under special direction from the architect. Not only do they learn to know one another much better, and have confidence in one another in their business relations, but more satisfactory business methods are thus possible through the fulness of their understanding. Nearly every one knows six months or a year before he is to begin building, and hence there is no good reason why this plan should not in most instances be carried out.

The spirit of domesticity is a dominant force in our time. The love of home is a sentiment high enough and strong enough to form the nucleus of great art.

The emotions which originate in family attachments and home life have a seriousness and delicacy which might belong to a Greek.

Great architecture has always been the expression of high sentiment. Greek architecture was the expression of an advanced intellectual condition. Gothic architecture was developed from great religious emotion and the exaltation of dawning intellectual freedom. The Renaissance was the result of research without the original impulse which belonged to the architecture of Greece, or that of the thirteenth century. But the architects of to-day have encumbered themselves with tradition. They have studied history, and neglected its philosophy. We are carrying with us the mere results of Greek and Mediæval art. True architecture must come from an original impulse. Our opportunity is in home life. Our architecture must spring from the family, and express all that is beautiful, tender, and ennobling in family love. It cannot be of temples, great cathedrals, or castles, because the sentiment which developed these is wanting. It must be of the family. It must relate to the love of men and women and children, youth and old age.

The world has never had a worthier motive for great art. The architecture of the thirteenth century was born full-strength out of the sentiment of that time.

Art in building contains less of vitality than any of the other arts, because it is the one art which rests solely on the sentiment of the past. To progress we must deal with the present. The thirteenth-century cathedral was perfectly suited to its practical demand. It was decorated through the mind of that time from the flora, and the imaginative traditions touched by the high sentiment of that period.

Art is not resuscitation: it is creation. The human heart is essentially the same that it has been in all great periods. All that is needed is a natural expression of a strong creative emotion.

The home must be suited to its practical requirements. It must be decorated through suggestions which come from our fields, as seen by the men of this time. We may draw from our intellectual resources, but we must always remember that our hearts beat in the nineteenth century.

"A VERY PRACTICAL BOOK."

— *Pittsburgh Telegraph.*

CONVENIENT HOUSES.

By LOUIS H. GIBSON, Architect,
Author of "Beautiful Houses"

CONTAINING a great variety of plans, photographic designs, and artistic exteriors and interiors of Ideal Homes, ranging in cost from $1,000 to $10,000. 8vo. $2.50.

This volume offers a practical solution of the vexed question of combining beauty with convenience in low-cost dwellings. It is a perfect revelation of what can be done at small expense.

Contents of the Book.

THE ARCHITECT AND THE HOUSEWIFE. Labor-saving devices, economy, and good construction, housekeeping operations, modern conveniences, plans that make extra work, misplaced houses, affectation in design, etc., etc.

A JOURNEY THROUGH THE HOUSE. Porch, vestibule, hall, long halls and square halls, parlor, sitting-room, dining-room, kitchens, a plan, fittings, dish-washing conveniences, sink, and tables, china-closet, pantry, dough-board, flour-bin, refrigerator arrangements, ventilation of kitchen, cellar, furnace, coal-bins, laundry, stairways, bedrooms, closets, gas-fixtures, bath-room, attic, plumbing, sewer connections, soil-pipe, water-closet, simplicity in plumbing, drain connections, grease-sink, heat and ventilation, the house and its beauty, artistic surroundings, mantels and grates, wood-carving, stained glass, quiet and light, external and internal design, etc., etc.

PLANS OF FIFTY CONVENIENT HOUSES. Evolution of a house-plan, respectable dimensions for a moderate price, a small pocket-book and a large idea, "We know what we want." What can be done for $1,600? One-story plans, side-hall plans, miscellaneous plans, varying costs, square plans, one-chimney plans, double houses, shingle houses, brick houses, etc.

PRACTICAL HOUSE-BUILDING FOR THE OWNER. Practical points, water, location, mason-work, foundations, flues, cisterns, laying brick, colored bricks, chimneys, hollow walls, grates, cut stone, terra-cotta, carpenter-work, height of stories, roof, floors, soft and hard wood floors, plastering, gas-piping, tin-work, gutters, painting, staining, oil-finishing, glazing, cost of a house, schedules of costs, what goes into a house, cost details, etc., etc.

BUSINESS POINTS IN BUILDING. Low-cost houses, method of making contracts, architects' estimates, building by the day, the safest plan, guarding against liens.

HOW TO PAY FOR A HOME. Building associations, purchase of a lot, etc., etc.

Press Notices.

The Advance.
"The house-builder and housekeeper has probably never had so admirable a pandect of the facts which he desires to know in planning for a future home as is now presented to him in 'Convenient Houses.' . . . The mechanical execution of the book, in illustration, printing, and binding, is tasteful and beautiful."

The Builder and Wood-Worker.
"In the section devoted to practical house-building, the author is at his best, writing with a clearness and simplicity of style that make his meaning entirely plain to the non-technical reader."

Christian Intelligencer.
"We heartily recommend the study of this book by those thinking of building a house for a home."

Living Church.
"We think a perusal of Mr. Gibson's book, and an examination of the plans given therein, would be of the greatest advantage to those interested in building 'convenient homes' for themselves or others."

New York Home Journal.
"The book seems to impart knowledge about everything concerning the house, from the sky-light to the coal-bin."

The Standard.
"It becomes clear, as one turns the pages, that this one architect, at all events, recognizes the fact that houses are built to live in, and that convenience is quite as much to be consulted as beauty, and a good deal more than showiness."

Boston Transcript.
"A volume of remarkable interest and value in these house-building days. . . . The illustrations are many and excellent."

American Agriculturist.
"To those of small incomes desirous of building a home for their families, the information contained in these last chapters alone is worth the price of the whole book."

Philadelphia Ledger.
"The most complete and practical of recent books on house-building."

Indianapolis News.
"Occupies a field of its own, touching upon points never before embraced in a book of the kind."

New York Tribune.
"A more thoroughly practical book is seldom printed."

For Sale by all Booksellers, or sent postpaid by the Publishers on receipt of price.

T. Y. CROWELL & CO., New York and Boston.

www.ingramcontent.com/pod-product-compliance
Lightning Source LLC
Chambersburg PA
CBHW020322240426
43673CB00039B/893